器官纤维化模型建立与植物提取物防治

Establishment of Organ Fibrosis Models and Prevention with Plant Extracts

张志刚
郭昌明 | 主编
高瑞峰

化学工业出版社
·北京·

内容简介

本书全面系统地介绍了器官纤维化及植物提取物在对抗纤维化中的潜在应用,涵盖器官纤维化的概念、发病机制以及各器官纤维化模型的建模方法,并且重点探讨了热点植物提取物在防治器官纤维化中的应用及作用机制。提供了关于纤维化机理和药物开发的详细信息,帮助读者理解纤维化的发展过程并探讨更有效的治疗方法。通过深入研究纤维化的发病机制和植物提取物的作用及其机理,旨在为读者提供有关器官纤维化的前沿研究进展及其应用前景。

本书可供从事兽医学和医学研究人员、临床医生、兽医工作者及药物开发人员参考,也可作为医学、药学、生物学领域的生产实践人员及高等院校相关专业师生的参考书。

图书在版编目(CIP)数据

器官纤维化模型建立与植物提取物防治 / 张志刚,郭昌明,高瑞峰主编. -- 北京:化学工业出版社,2025.3. -- ISBN 978-7-122-47574-9

I. Q946;R364.2

中国国家版本馆 CIP 数据核字第 2025AH7044 号

责任编辑:孙高洁　刘　军　冉海滢　　文字编辑:张春娥
责任校对:赵懿桐　　　　　　　　　　　装帧设计:王晓宇

出版发行:化学工业出版社
　　　　(北京市东城区青年湖南街 13 号　邮政编码 100011)
印　　装:北京天宇星印刷厂
710mm×1000mm　1/16　印张 8½　字数 210 千字
2025 年 5 月北京第 1 版第 1 次印刷

购书咨询:010-64518888　　　　　　　售后服务:010-64518899
网　　址:http://www.cip.com.cn
凡购买本书,如有缺损质量问题,本社销售中心负责调换。

定　价:80.00 元　　　　　　　　　　　版权所有　违者必究

本书编审人员名单

主　　编	张志刚	东北农业大学
	郭昌明	吉林大学
	高瑞峰	内蒙古农业大学
副 主 编	韩碧琦	东北农业大学
	李心慰	吉林大学
	李佳益	东北农业大学
	李小兵	云南农业大学
	黄宇翔	黑龙江省农业科学院
参编人员	李思雨	内蒙古农业大学
	孔　涛	河南科技大学
	杨大千	北京大学国际癌症研究院
	郭天龙	内蒙古农牧业科学院
	于美玲	广西大学
	潘志忠	松原职业技术学院
	杨　斌	内蒙古农牧业科学院
	吴晨晨	西北农林科技大学
	逯静静	黑龙江八一农垦大学
	纪淑珂	东北农业大学
	吕占军	东北农业大学
	杜希良	吉林大学
	薛江东	内蒙古民族大学
	朱言柱	吉林农业科技学院
	黄淑成	河南农业大学
	贺鹏飞	内蒙古农业大学
	宋玉祥	吉林大学
	刘　燕	内蒙古民族大学

张海莉　黑龙江省牡丹江市农业农村局
曹子凌　东北农业大学
雷　林　吉林大学
杨　旭　内蒙古锡林郭勒盟农牧局
刘运枫　东北农业大学
郭鑫宇　黑龙江省望奎县畜牧综合服务中心
王建国　西北农林科技大学
王　妍　西北农林科技大学
龙　淼　沈阳农业大学
李　鹏　沈阳农业大学
张　燚　沈阳农业大学
柳东芳　东北农业大学
宋凯薇　东北农业大学
李苗苗　东北农业大学
耿　洪　东北农业大学
马家瞳　东北农业大学

主　审　刘国文　吉林大学
　　　　王　哲　吉林大学

前言

纤维化是急性或慢性组织损伤时细胞外基质过度沉积和无序沉积引起的瘢痕和组织硬化，可发生于多种器官，如皮肤、肺脏、肝脏和肾脏等。其是许多疾病致残、致死的主要原因之一。然而，目前对器官纤维化的发病机制仍缺乏深入研究，几乎没有专门针对纤维化的治疗药物。而植物提取物具有来源天然且广泛，无毒副作用、低残留、无污染、可再生的独有优势。近年来，以植物提取物作为抗纤维化药物在纤维化的治疗方法研究中取得了诸多进展。由于与器官纤维化相关疾病的分子机制复杂，探寻通过充当拮抗剂或激动剂而调节信号转导和细胞生物活性的植物提取物用于治疗纤维化是一项重大挑战。

环境污染物暴露是导致器官纤维化相关疾病的重要因素之一，早在2010年本研究团队就开始关注环境污染物对人和动物的毒性作用及损伤机制。经过多年的探索和总结，我们在器官纤维化发生机制及抗纤维化植物提取物开发的研究中积累了一定经验，系列研究揭示了器官纤维化发生的新机制，证明了广泛分布的植物提取物通过靶向新机制而发挥抗纤维化作用，为开发安全有效的抗器官纤维化植物提取物制剂提供了科学依据。为推动我国植物提取物抗器官纤维化向更高层次发展，我们组织编写了本书。

本书共分三章，第一章简述了参与器官纤维化的细胞类型、常见的信号通路以及抗纤维化植物提取物的研究进展；第二章介绍了皮肤、肺脏、肝脏和肾脏等器官纤维化以及常用的各器官纤维化动物模型造模方法；第三章归纳了槲皮素、水飞蓟素和山柰酚等热点植物提取物抗器官纤维化的药理作用及其机制研究进展。

除整理国内外有关器官纤维化模型建立及热点植物提取物多部著作的前沿研究成果外，本书还紧密结合了本研究团队近年来关于植物提取物防治器官纤维化的"国家自然科学基金面上项目"研究成果。

由于水平和时间所限，虽然尽可能避免出错，但仍觉不足，书中疏漏之处在所难免，敬请各位同行不吝赐教和指正。

张志刚
2025年2月于哈尔滨

目录

第一章　器官纤维化与抗纤维化植物提取物概述 …………………… 001

第一节　器官纤维化概述 ………………………………………… 002
一、各器官纤维化概述 ………………………………………… 002
二、参与纤维化发生的细胞 …………………………………… 003

第二节　抗纤维化植物提取物概述 ……………………………… 007
一、抗纤维化植物提取物的分类 ……………………………… 007
二、天然植物提取物抗纤维化的机制 ………………………… 009

参考文献 …………………………………………………………… 010

第二章　器官纤维化及动物模型建立 ……………………………… 014

第一节　肝纤维化 ………………………………………………… 015
一、化学造模法 ………………………………………………… 015
二、乙醇造模法 ………………………………………………… 017
三、免疫造模法 ………………………………………………… 018
四、胆管结扎造模法 …………………………………………… 019

第二节　肾纤维化 ………………………………………………… 020
一、化学造模法 ………………………………………………… 020
二、手术造模法 ………………………………………………… 023

第三节　肺纤维化 ………………………………………………… 025
一、化学造模法 ………………………………………………… 025
二、物理造模法 ………………………………………………… 027

第四节　心肌纤维化 ……………………………………………… 028
一、化学造模法 ………………………………………………… 029
二、手术造模法 ………………………………………………… 031

第五节　胰腺纤维化	035
一、化学造模法	036
二、外科手术法	036
第六节　皮肤纤维化	036
一、化学造模法	037
二、物理造模法	037
参考文献	038

第三章　热点植物提取物防治器官纤维化的作用及机制　043

第一节　槲皮素	044
第二节　水飞蓟素	046
第三节　山奈酚	048
第四节　芹菜素	049
第五节　淫羊藿苷	052
第六节　黄芪甲苷	053
第七节　雷公藤红素	056
第八节　雷公藤内酯	059
第九节　青蒿琥酯	061
第十节　人参皂苷	063
第十一节　丹参酮ⅡA	070
第十二节　红景天苷	072
第十三节　姜黄素	075
第十四节　白藜芦醇	089
第十五节　木犀草素	099
参考文献	101

附录　119

英文缩略词对照表	119

第一章
器官纤维化与抗纤维化植物提取物概述

第一节 器官纤维化概述
第二节 抗纤维化植物提取物概述

第一节
器官纤维化概述

纤维化是创伤后伤口修复反应过程的重要组成部分，涉及多种细胞的相互作用，其主要特征是细胞外基质（extracellular matrix，ECM）的过量沉积，严重者导致组织和器官瘢痕化而出现高的发病率和死亡率。当组织出现损伤后，活化的成纤维细胞收缩性增强并分泌ECM，参与伤口愈合反应；损伤较轻或持续时间较短时，创伤愈合过程呈现为短暂的ECM积累，随后便迅速被机体清除；当组织损伤较为严重或持续时间较长时，体内的修复机制发生异常，使得组织正常结构遭到破坏，进而导致器官功能衰竭。大多数情况下，纤维化是由多种损伤因子驱动的实质性细胞死亡引起的，包括细胞坏死、细胞焦亡及铁死亡等。其具体过程为局部免疫细胞（主要是组织巨噬细胞）被激活进入损伤部位，释放高度活跃的细胞因子，进一步激活间充质细胞合成分泌ECM，增加促炎细胞因子、趋化因子和血管生成因子的释放。纤维化过程能够促进伤口愈合，但基质中的大量可溶性细胞因子使得组织修复失衡，组织结构随着胶原蛋白、弹性蛋白含量升高而发生改变，丰富的纤维间质形成了促纤维化的结缔组织增生的微环境，加快纤维化疾病进程。

一、各器官纤维化概述

肺纤维化是发展缓慢、肺组织遭到破坏并最终影响肺功能的间质性肺病，病理组织切片可观察到ECM沉积、炎细胞浸润及肺泡结构损坏。肺纤维化发生的诱因主要包括粉尘、化学品、外伤、免疫反应和病原微生物等，但其发病机制至今仍未完全阐明。肺纤维化的发展过程大致分为两个阶段，即早期组织持续性的炎症状态和后期组织修复异常引发的纤维化阶段[1]。

肝纤维化（liver fibrosis，LF）是在微生物或外源化学物刺激下，肝星状细胞（hepatic stellate cells，HSCs）被激活产生大量 ECM 并在肝沉积形成瘢痕组织的病理过程。在慢性肝损伤过程中，静止的 HSCs 被激活并改变为成纤维细胞形态，这些肝肌成纤维细胞是肝纤维化最下游的细胞效应因子，将直接导致 ECM 过度沉积和瘢痕形成。因此，HSCs 活化是引发 LF 的重要因素，也是大多数 LF 研究的主要靶点[2]。

肾纤维化是肾在致病因素作用下病理性修复的结果，是各种肾损伤的最终共同途径，表现为纤维化基质不受抑制地沉积，最终破坏器官结构，减少肾血液供应，扰乱器官功能。此外，纤维化还会降低组织正常的修复能力，导致肾衰竭[3]。

心肌纤维化是指心肌组织中正常心肌细胞减少，心肌成纤维细胞增生，ECM 过度生成和沉积的病理过程，是心脏损伤后心力衰竭、心律失常和猝死的主要危险因素[4]。

皮肤纤维化的特征是真皮层加厚以及一些附属物如毛囊、汗腺和皮肤血管的阻塞，是系统性硬化症的典型特征[5]。

胰腺纤维化是由胰腺星状细胞（pancreatic stellate cells，PSCs）激活引起的多种慢性胰腺炎的共同组织病理学特征，但其靶向干预策略迄今为止还不明确[6]。

二、参与纤维化发生的细胞

纤维化是多种细胞之间相互作用的结果。目前已经对特发性肺间质纤维化（idiopathic pulmonary fibrosis，IPF）、LF、肾纤维化和系统性硬化症等纤维化的细胞图进行了充分的研究。这些研究证实了上皮细胞、内皮细胞、免疫细胞和成纤维细胞在纤维化中的关键作用，并确定了参与病理进展的一些新细胞类型。

1. 上皮细胞

上皮细胞（包括基底细胞、分泌细胞、俱乐部细胞、纤毛细胞和杯状细胞）是许多器官维持组织稳态的关键细胞[7]。在纤维化过程中，慢性损伤导致上皮细胞凋亡，从而破坏上皮结构，促进功能失调性修复和成纤维细胞的致病性激活[8]。此外，上皮间质转化（epithelial-mesenchymal transition，EMT）被认为是肌成纤维细胞（myofibroblast，MF）的重要来源。病理状态下的 EMT 可导致正常上皮细胞减少，破坏组织的正常结构，促进胶原纤维的产生[9]。研究表明，上皮细胞（如肺泡上皮细胞、杯状细胞、纤毛细胞和俱乐部细胞）对肺纤维化的发生至关重要[10-11]。肺泡上皮细胞包括肺泡Ⅰ型上皮（alveolar type Ⅰ epithelial，AT1）细胞和肺泡Ⅱ型上皮（alveolar type Ⅱ epithelial，AT2）细胞，是肺组织中主要的上皮细胞，维持着肺泡壁的完整性。当损伤导致 AT1 细胞死亡时，AT2 细胞增殖并分化为 AT1 细胞，从而维持肺泡的正常结构[12]。在肺中发现了一种新的上皮细胞亚群 Axin2 AT2 细胞，具有祖细胞和上皮细胞的特性，并调节肺泡再生[13-14]。

在气道上皮中发现了一组新的高表达囊性纤维化穿膜传导调节蛋白（cystic fibrosis transmembrane conductance regulator，CFTR）的上皮细胞，命名为离子细胞。CFTR 最重要的功能之一是调节氯离子通道[15]。上皮细胞 CFTR 基因突变导致气道上皮氯离子通道缺陷，启动囊性纤维化的发生[16]。此外，气道中 CFTR 的缺乏会增加钠通道活性和钠的高吸收，提示 CFTR 可能参与了钠转运[17]。胰腺和肝脏上皮细胞的功能改变也受到 CFTR 突变的影响。在正常肝脏中，CFTR 与胆管细胞顶部的氯离子通道协同作用，为胆汁水化提供动力[18]。CFTR 功能受损导致胆管黏膜增生和梗阻。随后的胆盐蓄积导致了肝细胞损伤、炎症和门静脉纤维化[19]。

2. 内皮细胞

内皮细胞是血管的主要组成部分。内皮细胞受损导致血液与组织间物质交换异常，进而导致代谢紊乱。此外，在纤维化组织中，由于需要更多血液营养的成纤维细胞大量增殖，可能会诱导血管生成出现异常。研究表明，不同纤维化组织的内皮细胞也可能具有特定的功能。两个新的内皮细胞亚型，即原生质膜囊泡相关蛋白（plasmalemma vesicle associated protein，PLVAP）内皮细胞和非典型趋化因子受体 1（atypical chemokine receptor 1，ACKR1）内皮细胞被发现在肝硬化患者肝组织中可以促进白细胞迁移。在肺组织中，单细胞测序鉴定出 4 组内皮细胞，包括毛细血管内皮细胞 A 和毛细血管内皮细胞 B、静脉内皮细胞和动脉内皮细胞。胶原蛋白 15 α1（collagen 15A1，

COL15A1)基因高表达所识别出的第五类内皮细胞,位于细支气管和纤维灶,参与 ECM 的产生。

3. 免疫细胞

免疫系统异常可能是纤维化的早期事件[20]。T 淋巴细胞、巨噬细胞、树突状细胞(dendritic cells,DCs)、粒细胞和肥大细胞等免疫细胞参与了纤维化进程[21]。这些活化的免疫细胞高表达调节炎症和纤维化的因子,促进了纤维母细胞的活化。在 IPF 患者中,T 淋巴细胞,包括 $CD4^+$ T 细胞、$CD8^+$ T 细胞和 CD8 效应细胞增加。IPF 患者中的 T 淋巴细胞的干扰素-γ 信号转导显著改变,而系统性硬化症患者中的 T 淋巴细胞的白细胞介素-6(IL-6)信号显著激活[22]。在肝组织中,细胞毒性 T 细胞的表达增加,$CD4^+$ T 细胞的失活可以诱导纤维化[23]。

巨噬细胞是在纤维化疾病中介导炎症和纤维化的关键细胞。在肝硬化患者的组织中鉴定出了七种巨噬细胞亚群,包括库普弗细胞(肝脏中的驻留巨噬细胞)和表达 CD9、髓系细胞触发受体 2(triggering receptor expressed on myeloid cells 2,TREM2)巨噬细胞。伪时序分析显示,TREM2 CD9 巨噬细胞起源于单核细胞,并增加了 HSCs 的胶原蛋白表达。在肺纤维化中,鉴定出了 18 种免疫细胞类型,并确定了组织驻留巨噬细胞、促纤维化表型巨噬细胞和炎症巨噬细胞的表型。肺的组织驻留巨噬细胞主要是肺泡巨噬细胞(alveolar macrophages,AMs)[24]。AMs 紧密附着于肺泡上皮并暴露于外界环境。可吸入颗粒和其他因素直接导致 AMs 的死亡。活化的 AMs 分泌炎症介质来激活炎症反应,并提高纤维化因子的表达,以促进肺纤维化[25-26]。肺纤维化小鼠模型中还发现了与纤维母细胞相邻的糖醛酸结合 Ig 样凝集素 F(SiglecF)C-X3-C 基序趋化因子受体 1(CX3CR1)巨噬细胞,通过释放血小板衍生生长因子(platelet-derived growth factors,PDGFs)促进纤维母细胞的增殖和活化,从而促进纤维化[27]。

4. 成纤维细胞

在许多纤维化疾病中,成纤维细胞的分化是关键的细胞事件,其向具有分泌、收缩和 ECM 合成功能的 MF 转变。有研究证实,MF 在不同器官和不同纤维化病理状态下，表现出多样的基因表达谱[28-29]。在肺组织中，成纤维细胞的分化途径在正常状态和纤维化进程中有所不同。在正常情况下，间充质祖细胞分化为脂肪成纤维细胞和 COL14A1 基质成纤维细胞，而后者再进一步分化为 MF 和 COL13A1 基质成纤维细胞。然而，在肺纤维化中，间充质祖细胞可分化为脂肪成纤维细胞、$PDGFR\beta^{++}$ hi 亚型成纤维细胞、COL14A1 基质成纤维细胞、MF 和 COL13A1 基质成纤维细胞[30]。在肝脏纤维化中，HSCs 是纤维细胞的主要类型，其以星状形态为特征。HSCs 的分化过程包括静止状态的解除、炎症促进、迁移和 ECM 的产生。

纤维化过程中，免疫细胞、EMT 和内皮间质转化（endothelial-to-mesenchymal transition，EndMT）被认为是 MF 数量和活化增加的主要贡献因素[31-32]。抑制 MF 的增殖和活化一直是纤维化治疗的关键挑战。然而，在纤维化过程中，MF 可能获得凋亡抵抗力，阻碍了程序性细胞死亡机制的实施[33]。减少 MF 数量的治疗方法的效果有限，此外，MF 的过度活化通常是实质细胞（如上皮细胞、心肌细胞和内皮细胞）死亡的代偿反应。因此，通过减少实质细胞和其他相关细胞的死亡或调节其活性，以间接抑制 MF 的活化可能是一种更有效的治疗策略。

在肝脏纤维化过程中，HSCs 与其他细胞的相互作用机制是复杂的。在正常肝脏中，肝窦内皮细胞的分化维持可导致 HSCs 处于静息状态，并促进纤维化的逆转[34-35]。然而，在纤维化过程中，凋亡肝细胞的增加引发炎症反应并激活肝脏内的巨噬细胞[36]。来自库普弗细胞、肝细胞、B 淋巴细胞和 T 淋巴细胞的细胞外因素 EMT 进一步调节 HSCs 的活化[37]。通过调节维甲酸诱导基因 1/天然杀伤细胞组 2D 相关的凋亡和与肿瘤坏死因子相关的诱导凋亡配体，自然杀伤细胞可杀伤活化的 HSCs[38]。慢性肝损伤导致持续的 HSCs 活化，进而促进 ECM 的积累和组织结构的重塑，从而推动 LF 的进行[39]。

在肺部，肺泡上皮细胞的急性损伤可导致上皮细胞减少、肺泡结构破坏和炎症因子释放，从而激活免疫细胞。这些活化的炎症细胞和受损的上皮细胞增加了肿瘤坏死因子-α（tumor necrosis factor α，TNF-α）、白细胞介素-1β（interleukin-1β，IL-1β）、白细胞介素-6 和转化生长因子-β（transforming growth factor-β，TGF-β）等细胞因子的表达[40-41]。在初始炎症事件之后，肺成纤维细胞通过上调 PDGFs、成纤维细胞生长因子和血管内皮生长因子（vascular endothelial growth factors，VEGFs）等纤维化相关细胞因子，激活为 MF。此外，上皮细胞通过经历 EMT 的过程，也可能增加 MF 的数量。慢性活化的 MF 产生胶原蛋白、纤维连接蛋白和蛋白多糖等 ECM 成分，推动肺纤维化的进展[42]。

第二节
抗纤维化植物提取物概述

植物提取物是多种化学物质的丰富来源，它可以抵御生物和非生物威胁，比如病原体、辐射、重金属或气候变化。已知植物次生代谢产物包括酚类化合物、生物碱、皂苷等多种类别。这些物质在人类身体中具有广泛的生物效应，包括抗氧化、抗炎、抗癌、抗菌、抗病毒、抗肥胖、抗糖尿病、抗骨质疏松以及心脏保护或神经保护等性质。

一、抗纤维化植物提取物的分类

1. 酚类化合物

酚类化合物的特点是至少含有一个酚羟基，这一类化合物可以根据结构分为简单酚类、单宁、香豆素、类黄酮、木脂素和木酚素。简单酚类包含一个 C_6 苯环，连接有一个或多个羟基、醛基或羧基。单宁可以与蛋白质和碳水化合物形成复合物。香豆素是苯并-α-吡喃酮（内酯）的衍生物，由 C_6-C_3 碳骨架组成，可以以游离形式存在，也可以与糖结合成为糖苷。类黄酮是数量最多的亚类，它们具有 C_6-C_3-C_6 基本碳骨架，形成了闭合的吡喃 C 环作为中心碳链，两个苯环 A 和 B 连接在 C 环的 2 位、3 位或 4 位。异黄酮和新黄酮是类黄酮的一类，其中 B 环分别连接在 C 环的 3 位和 4 位。B 环连接在 C 环的 2 位的类黄酮，可以根据 C 环的结构特征进一步分为几类，包括黄酮、黄酮醇、二氢黄酮、二氢黄酮醇、黄烷-3-醇（儿茶素属于此类）、花青素等。另外两个常见的类别是木脂素和木酚素，它们是苯并-γ-吡喃的衍生物。天然存在的香豆素由具有甲基或烷基取代基的 C_6-C_3 基本碳骨架组成，5 位和 7 位有羟基或烷氧基取代基，而木脂素由 C_6-C_3-C_6 碳骨架组成，带有各种取代基。此外，木脂素主要存在于植物的木质部和树脂中，而木酚素（C_6-C_3）是由两个苯丙素衍生物二聚化形成。这些化合物大多具有清除自由基的性质，这是由它们分子中羟基（—OH）和甲氧基（—OCH$_3$）的存在所决定的。已知许多植物是酚类化合物的丰富来源，包括菊科、蔷薇科和唇形科。大量证据表明，酚类化合物对人类成纤维细胞具有显著的抗氧化作用，能提高成纤维细胞的存活能力和活动性[43]。

2. 生物碱

生物碱主要来源于植物，部分来源于真菌和动物。它们是由超过 12000 种不同的结构构成的一个非常庞大且多样化的次生化合物群。传统定义认为生物碱是低分子量化合物，其含有异环氮，来源于氨基酸，并且在水中表现为碱性反应。然而，进一步的植物次生代谢产物，例如麻黄碱来自麻黄（麻黄科）和秋水仙碱来自秋水仙（百合科），它们都含有非异环氮原子，也被视为生物碱。另一个生物碱的定义侧重于许多生物碱所产生的强烈药理效应。一个重要的例子是来自太平洋红豆杉（红豆杉科）的紫杉醇，尽管其核心骨架源自二萜，分子中唯一的非碱性氮来自苯基异丝氨酸。生物碱在哺乳动物中表现出的显著效应自古以来就被利用。《埃伯斯草纸》收录了一批埃及药用植物和配方，可追溯至公元前约 1550 年。它列举了茄科植物，其中含有阿托品和东莨菪碱，如埃及颠茄（毒曼陀罗）用于牙痛[44]。一些生物碱的抗 LF 作用已逐渐被发现。如今，随着肝脏疾病受到越来越多的关注，生物碱的抗肝纤维化作用也被广泛研究[45]。

3. 萜类化合物

萜类化合物是烯烃化合物，每种化合物的分子式都是 $(C_5H_8)_n$，即异戊二烯的整数倍。一些天然萜烯在自然界中广泛存在，并具有抗肝纤维化活性[46]。许多中药天然产物，如甘草甜素、柴胡、三七和人参都含有萜类化合物。

二、天然植物提取物抗纤维化的机制

1. 抑制炎症

有研究发现，刺五加生物碱 A 及其乙酰化衍生物对 LF 具有有益作用，其机制可能与抑制炎症反应有关。氧化苦参碱能有效减弱四氯化碳（carbon tetrachloride，CCl_4）诱导的 LF，这种作用可能是由于 Toll 样受体 4（toll-like receptor 4，TLR4）依赖性炎症和 TGF-β1 信号通路的调节。双黄连多糖可有效拮抗二甲基亚硝胺（N-nitrosodimethylamine，NDMA）活性，其抗纤维化机制与调节功能关键酶、改善代谢功能和抑制组织中的炎症反应有关。黄芩苷抑制 TNF-α 蛋白的表达，对 CCl_4 诱导 LF 大鼠有保护作用。葛根素能够逆转 CCl_4 诱导的肝毒性，通过调节血清酶活力并减弱 TNF-α/NF-κB（nuclear factor-kappa B，NF-κB）通路的抗炎反应来发挥作用。青蒿琥酯可通过抑制大鼠 LPS/TLR4/NF-κB 信号通路缓解多种致病因素和炎症诱导的纤维化[47]。

2. 抑制 ECM 的合成

纤维化是 ECM 过度沉积的不良结果，胶原是 ECM 的主要成分。辣椒素对 DMN 和 TGF-β1 诱导的大鼠 LF 具有抑制作用，辣椒素可抑制 DMN 诱导的肝毒性、NF-κB 活化和胶原积累。辣椒素通过 PPAR-γ 激活抑制 TGF-β1/Smad 通路，从而改善 LF。粉防己碱是从中草药粉防己中分离得到的一种生物碱。有研究评估了粉防己碱对 LF 的体外和体内影响，发现粉防己碱或水飞蓟素处理可显著降低 LF 大鼠的肝胶原含量。此外，粉防己碱还抑制 TGF-β1 诱导的 HSC-T6 细胞中 α-平滑肌肌动蛋白（alpha-smooth muscle actin，α-SMA）分泌和胶原沉积。橙皮素是柑橘类水果的天然成分，作为 TGF-β 抑制剂，被证明对损伤和各种癌症具有有益的抗炎作用。从柑橘皮中提取橙皮素可以通过抑制 TGF-β1/Smad 通路介导的 ECM 增加和细胞凋亡来阻止胆管结扎（bile duct ligation，BDL）诱导的 LF[48]。

参考文献

[1] Huang D Y, Li Y H, Liu Y. Tacrolimus and the treatment of pulmonary fibrosis[J]. Am J Respir Crit Care Med, 2023.

[2] Hammerich L, Tacke F. Hepatic inflammatory responses in liver fibrosis [J]. Nat Rev Gastroenterol Hepatol, 2023, 20(10): 633-646.

[3] Klinkhammer B M, Boor P. Kidney fibrosis: Emerging diagnostic and therapeutic strategies[J]. Mol Aspects Med, 2023, 93: 101206.

[4] Stachowicz A, Sadiq A, Walker B, et al. Treatment of human cardiac fibroblasts with the protein arginine deiminase inhibitor BB-Cl-amidine activates the Nrf2/HO-1 signaling pathway[J]. Biomed Pharmacotherapy, 2023, 167: 115443.

[5] Talbott H E, Mascharak S, Griffin M, et al. Wound healing, fibroblast heterogeneity, and fibrosis [J]. Cell Stem Cell, 2022, 29(8): 1161-1180.

[6] Yang X G, Chen J, Wang J, et al. Very-low-density lipoprotein receptor-enhanced lipid metabolism in pancreatic stellate cells promotes pancreatic fibrosis[J]. Immunity, 2022, 55(7): 1185-1199.

[7] Carraro G, Langerman J, Sabri S, et al. Transcriptional analysis of cystic fibrosis airways at single-cell resolution reveals altered epithelial cell states and composition[J]. Nat Med, 2021, 27(5): 806-814.

[8] Parimon T, Yao C, Stripp B R, et al. Alveolar epithelial type II cells as drivers of lung fibrosis in idiopathic pulmonary fibrosis[J]. Int J Mol Sci, 2020, 21(7): 2269.

[9] Yao L D, Conforti F, Hill C, et al. Paracrine signalling during ZEB1-mediated epithelial-mesenchymal transition augments local myofibroblast differentiation in lung fibrosis[J]. Cell Death Differ, 2019, 26(5): 943-957.

[10] Reyfman P A, Walter J M, Joshi N, et al. Single-cell transcriptomic analysisof human lung provides insights into the pathobiology of pulmonary fibrosis[J]. Am J Respir Crit Care Med, 2019, 199(12): 1517-1536.

[11] Xu Y, Mizuno T, Sridharan A, et al. Single-cell RNA sequencing identifies diverse roles of epithelial cells in idiopathic pulmonary fibrosis[J]. JCI Insight, 2016, 1(20): e90558.

[12] Barkauskas C E, Cronce M J, Rackley C R, et al. Type 2 alveolar cells are stem cells in adult lung [J]. J Clin Invest, 2013, 123(7): 3025-3036.

[13] Zacharias W J, Frank D B, Zepp J A, et al. Regeneration of the lung alveolus by an evolutionarily conserved epithelial progenitor[J]. Nature, 2018, 555(7695): 251-255.

[14] Zepp J A, Zacharias W J, Frank D B, et al. Distinct mesenchymal lineages and niches promote epithelial self-renewal and myofibrogenesis in the lung[J]. Cell, 2017, 170(6): 1134-1148.

[15] Montoro D T, Haber A L, Biton M, et al. A revised airway epithelial hierarchy includes CFTR-expressing ionocytes[J]. Nature, 2018, 560(7718): 319-324.

[16] Ratjen F, Döring G. Cystic fibrosis[J]. Lancet, 2003, 361(9358): 681-689.

[17] Olivier A K, Gibson-Corley K N, Meyerholz D K. Animal models of gastrointestinal and liver diseases. Animal models of cystic fibrosis: gastrointestinal, pancreatic, and hepatobiliary disease and pathophysiology[J]. Am J Physiol Gastrointest Liver Physiol, 2015, 308(6): G459-G471.

[18] Ledder O, Haller W, Couper R T, et al. Cystic fibrosis: An update for clinicians. Part 2: hepatobiliary and pancreatic manifestations [J]. J Gastroenterol Hepatol, 2014, 29 (12): 1954-1962.

[19] Feranchak A P, Sokol R J. Cholangiocyte biology and cystic fibrosis liver disease[J]. Semin Liver Dis, 2001, 21(4): 471-488.

[20] Frantz C, Cauvet A, Durand A, et al. Driving role of interleukin-2-related regulatory CD4+ T cell deficiency in the development of lung fibrosis and vascular remodeling in a mouse model of systemic sclerosis[J]. Arthritis Rheumatol, 2022, 74(8): 1387-1398.

[21] Valenzi E, Tabib T, Papazoglou A, et al. Disparate interferon signaling and shared aberrant basaloid cells in single-cell profiling of idiopathic pulmonary fibrosis and systemic sclerosis-associated interstitial lung disease[J]. Front Immunol, 2021, 12: 595811.

[22] Serezani A P M, Pascoalino B D, Bazzano J M R, et al. Multiplatform single-cell analysis identifies immune cell types enhanced in pulmonary fibrosis[J]. Am J Respir Cell Mol Biol, 2022, 67(1): 50-60.

[23] Liang Q, Hu Y D, Zhang M N, et al. The T cell receptor immune repertoire protects the liver from reconsitution[J]. Front Immunol, 2020, 11: 584979.

[24] Reyfman P A, Walter J M, Joshi N, et al. Single-cell transcriptomic analysisof human lung provides insights into the pathobiology of pulmonary fibrosis[J]. Am J Respir Crit Care Med, 2019, 199(12): 1517-1536.

[25] Prasse A, Pechkovsky D V, Toews G B, et al. A vicious circle of alveolar macrophages and fibroblasts perpetuates pulmonary fibrosis via CCL18[J]. Am J Respir Crit Care Med, 2006, 173(7): 781-792.

[26] Zhang W, Ohno S, Steer B, et al. S100a4 is secreted by alternatively activated alveolar macrophages and promotes activation of lung fibroblasts in pulmonary fibrosis[J]. Front Immunol, 2018, 9: 1216.

[27] Aran D, Looney A P, Liu L, et al. Reference-based analysis of lung single-cell sequencing reveals a transitional profibrotic macrophage[J]. Nat Immunol, 2019, 20(2): 163-172.

[28] Farbehi N, Patrick R, Dorison A, et al. Single-cell expression profiling reveals dynamic flux of cardiac stromal, vascular and immune cells in health and injury[J]. Elife, 2019, 8: e43882.

[29] Tabib T, Morse C, Wang T, et al. SFRP2/DPP4 and FMO1/LSP1 define major fibroblast populations in human skin[J]. J Invest Dermatol, 2018, 138(4): 802-810.

[30] Xie T, Wang Y Z, Deng N, et al. Single-cell deconvolution of fibroblast heterogeneity in mouse pulmonary fibrosis[J]. Cell Rep, 2018, 22(13): 3625-3640.

[31] Yuan Q, Tan R J, Liu Y H. Myofibroblast in kidney fibrosis: Origin, activation, and regulation [J]. Adv Exp Med Biol, 2019, 1165: 253-283.

[32] Zeisberg E M, Tarnavski O, Zeisberg M, et al. Endothelial-to-mesenchymal transition contributes to cardiac fibrosis[J]. Nat Med, 2007, 13(8): 952-961.

[33] Kato K, Logsdon N J, Shin Y J, et al. Impaired myofibroblast dedifferentiation contributes to nonresolving fibrosis in aging[J]. Am J Respir Cell Mol Biol, 2020, 62(5): 633-644.

[34] Xie G H, Wang X D, Wang L, et al. Role of differentiation of liver sinusoidal endothelial cells in progression and regression of hepatic fibrosis in rats[J]. Gastroenterology, 2012, 142(4): 918-927.

[35] Deleve L D, Wang X D, Guo Y M. Sinusoidal endothelial cells prevent rat stellate cell activation and promote reversion to quiescence[J]. Hepatology, 2008, 48(3): 920-930.

[36] Seki E, Schwabe R F. Hepatic inflammation and fibrosis: functional links and key pathways[J]. Hepatology, 2015, 61(3): 1066-1079.

[37] Tsuchida T, Friedman S L. Mechanisms of hepatic stellate cell activation[J]. Nat Rev Gastroenterol Hepatol, 2017, 14(7): 397-411.

[38] Jeong W I, Park O, Suh Y G, et al. Suppression of innate immunity (natural killer cell/interferon-γ) in the advanced stages of liver fibrosis in mice[J]. Hepatology, 2011, 53(4): 1342-1351.

[39] Rosselli M, Lotersztajn S, Vizzutti F, et al. The metabolic syndrome and chronic liver disease [J]. Curr Pharm Des, 2014, 20(31): 5010-5024.

[40] Ramachandran P, Iredale J P, Fallowfield J A. Resolution of liver fibrosis: basic mechanisms and clinical relevance[J]. Semin Liver Dis, 2015, 35(2): 119-131.

[41] Zhang Y, Lee T C, Guillemin B, et al. Enhanced IL-1 beta and tumor necrosis factor-alpha release and messenger RNA expression in macrophages from idiopathic pulmonary fibrosis or after asbestos exposure[J]. J Immunol, 1993, 150(9): 4188-4196.

[42] Distler J H W, Györfi A H, Ramanujam M, et al. Shared and distinct mechanisms of fibrosis[J]. Nat Rev Rheumatol, 2019, 15(12): 705-730.

[43] Merecz-Sadowska A, Sitarek P, Kucharska E, et al. Antioxidant properties of plant-derived phenolic compounds and their effect on skin fibroblast cells[J]. Antioxidants (Basel), 2021, 10 (5): 726.

[44] Schläger S, Dräger B. Exploiting plant alkaloids[J]. Curr Opin Biotechnol, 2016, 37: 155-164.

[45] Wang Q, Dai X F, Yang W Z, et al. Caffeine protects against alcohol-induced liver fibrosis by dampening the cAMP/PKA/CREB pathway in rat hepatic stellate cells Int [J].

Immunopharmacol, 2015, 25(2): 340-352.

[46] Hu Z T, Qin F, Gao S L, et al. Paeoniflorin exerts protective effect on radiation-induced hepatic fibrosis in rats via TGF-β1/Smads signaling pathway[J]. Am J Transl Res, 2018, 10(3): 1012-1021.

[47] Shan L, Liu Z N, Ci L L, et al. Research progress on the anti-hepatic fibrosis action and mechanism of natural products[J]. Int Immunopharmacol, 2019, 75: 105765.

[48] Kong R, Wang N, Luo H, et al. Hesperetin mitigates bile duct ligation-induced liver fibrosis by inhibiting extracellular matrix and cell apoptosis via the TGF-β1/Smad pathway[J]. Curr Mol Med, 2018, 18(1): 15-24.

CHAPTER
02

第二章
器官纤维化及动物模型建立

第一节　肝纤维化
第二节　肾纤维化
第三节　肺纤维化
第四节　心肌纤维化
第五节　胰腺纤维化
第六节　皮肤纤维化

第一节
肝纤维化

肝纤维化（LF）的概念是20世纪50年代由德国科学家提出的，其被认为是癌前状态，与增加原发性肝癌相对死亡率密切相关，占肝癌患者死亡率的一半。因营养代谢和病毒感染等致慢性肝病的各种原因都可引发LF，在其发病过程中首先出现肝损伤导致ECM积聚在肝中，随后导致LF和肝硬化。LF的发生发展机制是由多种细胞、蛋白和各种促纤维化细胞因子等因素共同构成，其中HSCs是主要的细胞类型。

一、化学造模法

1. 四氯化碳诱导肝纤维化

建立CCl_4诱导肝纤维化模型是一种经典的造模方法，因其具有造模成功率高、方法简便、费用低、时间短等优势，成为临床基础研究中肝纤维化造模的首选方法。CCl_4由于其可引起肝脏损伤及纤维化而被广泛用于诱导肝纤维化动物模型。其机制是直接溶解肝细胞的细胞膜，经肝细胞细胞色素P450依赖性混合功能氧化酶的代谢生成高反应性的·CCl_3，导致小叶中央肝细胞坏死，触发基质在中央静脉周围沉积并逐步与相邻的中央静脉形成桥接，并且几乎不发生胆管反应。制备方法主要有以下四种：

（1）单纯CCl_4诱导法 该方法按0.5～2mL/kg体重腹腔注射1∶1的CCl_4橄榄油溶液，6～9周后可以导致LF[1]。

（2）CCl_4联合苯巴比妥制备法 该方法用含0.35g/L的苯巴比妥钠的蒸馏水作为大鼠的唯一饮用水，饮用1周后，按0.05mL/kg体重腹腔注射40% CCl_4中性菜籽油溶液4周，每周2次，同时以10%的乙醇溶液为其唯一饮用水；第5周开始改为0.05mL/kg体重腹腔注射50% CCl_4中性菜籽油

4周,每周2次,以50%的乙醇溶液为其唯一饮用水[2]。

(3) CCl_4 复合法　该方法系采用高脂低蛋白食物(以含0.5%胆固醇的玉米面或实验第1、2周以含20%胆固醇的猪油为饲料),以30%的酒精为唯一饮用水,皮下注射CCl_4(第1次按0.05mL/kg体重注射,以后每隔3天按0.03mL/kg体重皮下注射40%的CCl_4猪油油剂,4周后即可形成LF)[3]。

(4) AAF/CCl_4联合制备法　该方法先给实验动物(如SD大鼠)0.02%颗粒状的乙酰氨基芴(acetylaminofluorene,AAF)混于饲料中喂养5天,后以1:1稀释的CCl_4橄榄油混合液按2mL/kg体重腹腔注射1周,每周2次,从第2周开始CCl_4剂量改为1mL/kg体重并停止加入AAF,共6周即可形成LF[4]。

2.二甲基亚硝胺诱导肝纤维化

NDMA也叫亚硝基二甲胺,其分子式为$C_2H_6N_2O$或$(CH_3)_2NNO$,具有巨大的肝毒性,对基因结构及免疫功能也具有一定的损伤作用,其主要作用机制是使核酸、蛋白质发生甲基化,大剂量使用NDMA易导致动物死亡,多次小剂量腹腔注射、灌胃及吸入可以导致肝细胞坏死。NDMA在体内通过肝微粒体主要代谢为乙醛。乙醛是一种高反应活性化学物,可与核酸、蛋白质等重要的生命物质结合并使其发生甲基化导致肝细胞坏死、ECM进行性增加,进而出现LF。

目前常用的造模方法是,通过给实验动物腹腔注射NDMA(5~10mg/kg体重),每周连续3天,4~8周后即可成功诱导出肝纤维化模型。高雅等给予大鼠1.6mL/kg NDMA溶液腹腔注射,一周3次,6周时大鼠出现明显的LF,即肝小叶消失、假小叶形成[5];王晶晶等给予大鼠腹腔注射0.5% NDMA生理盐水稀释液,剂量为2mL/kg体重,一周1~3次,共注射4周[6];李伟伟等通过腹腔注射0.5% NDMA溶液(2mL/kg体重),每2天注射1次,连续4周,可得到稳定的肝纤维化模型[7]。与CCl_4法相比,NDMA法具有周期短、方法简单的优点,停止给药后LF不易自行恢复,但是NDMA为致癌物质,其除了腹腔注射外,还可通过呼吸道、皮肤等途径产生毒性作用,因此实验操作者在实验时应格外小心谨慎。

3. 硫代乙酰胺诱导肝纤维化

硫代乙酰胺（thioacetamide，TAA）分子式为 CH_3CSNH_2，其诱导的急性肝损伤机制是 TAA 进入细胞内，被细胞色素氧化酶 P450 氧化后生成自由基，自由基与肝细胞膜脂质结合，诱发脂质过氧化破坏肝细胞膜，造成肝细胞的变性坏死。

主要的造模方法有灌胃、腹腔注射、皮下注射等，最常用的是皮下注射。秦冬梅等使用灭菌注射用水将 TAA 配制成质量浓度为 40mg/mL 的溶液，对大鼠进行腹腔注射，一周 3 次，8 周时肝纤维化模型造模成功[8]；薛改等按 160mg/kg，对大鼠隔天灌胃浓度为 3％ 的 TAA，每周称量体重 1 次，严格按照体重变化调整 TAA 的灌胃量，1 周内体重下降 5％ 以内者维持原用量，1 周内体重下降 5％～10％ 者用量减少为原用量的 1/2，1 周内体重下降 10％ 以上者停止使用以避免死亡，结果显示该方法可以降低大鼠的死亡率[9]；张帅等对大鼠皮下注射 0.03％ 的 TAA，剂量为 200mg/kg，一周 2 次，8 周时肝纤维化模型造模成功[10]。综上所述，TAA 造模时间较短，肝纤维化模型成功率高且相对稳定、死亡率低，可广泛应用于临床基础研究。但其具有易挥发、毒性大的特点，需要注意加强个人防护。

二、乙醇造模法

乙醇（俗称酒精）分子式为 C_2H_5OH，进入人体内的酒精有 90％ 在肝脏中代谢，其肝损伤机制是酒精进入肝脏细胞，在脱氢酶和微粒体乙醇氧化酶系的作用下转化为乙醛，进而转变为乙酸，促使还原型辅酶Ⅰ与氧化型辅酶Ⅰ比值升高，抑制线粒体三羧酸循环，使肝内脂肪酸代谢发生障碍，氧化减弱，使中性脂肪堆积于肝细胞中，从而引起肝细胞的炎症、变性、坏死，诱发胶原纤维堆积导致 LF。酒精性肝纤维化的模型主要有以下几种：

（1）Lieber-DeCarli 模型　让啮齿类动物自由饮用含有乙醇的液体饲料（唯一的饮食来源），其中乙醇可替代碳水化合物产生的 50％ 的热量，并逐步使实验动物摄入乙醇所产生的热量达到总摄入热量的 36％[11]。但是，该模型可能存在形成脂肪肝但不出现 LF 的风险，同时实验动物的厌酒诱导成功率低，给造模带来了很大的挑战。

（2）Tsukamoto-French模型　对实验动物采用持续胃内灌注含酒精的高脂流质饮食，从而维持一个相对较高的酒精血药浓度，可对肝脏造成损伤，进而诱导LF[12]。随后有研究对该法进行了改进，即给实验大鼠胃内持续灌注含有60%的乙醇（1mL/kg体重）、橄榄油（2mL/kg体重）和吡唑（25mg/kg体重），连续9周[13]。尽管该法可在短期内诱导LF，但LF程度较轻，不易发展为肝硬化，且实验技术要求较高、成本较高，推广应用尚需进一步改进，以使LF程度加重并降低造模成本。

（3）其他　还通过31.5%酒精溶液（95%酒精与水混合制成）灌胃的方法制备酒精性肝纤维化模型[14]。有学者采用体积分数为50%的酒精诱导肝纤维化模型，16周时成功复制酒精性肝纤维化模型，证明此法操作简单、经济低廉。李丰衣等采用"白酒-吡唑-植物油"混合液灌胃法制备乙醇肝纤维化动物模型，即每天给予每只大鼠以4.8g/kg白酒、24mg/kg吡唑、2mL/kg玉米油的混合液灌胃，灌胃量随体重的变化进行调整，同时喂养高脂饲料（基础饲料：胆固醇：猪油＝79：1：20），于16周时发现大鼠轻度酒精性LF形成[15]。

三、免疫造模法

自身免疫性LF的发病机制主要以遗传因素为主，其他因素是在遗传易感性基础上，机体免疫耐受机制破坏，从而产生针对肝脏自身抗原的免疫反应，诱发肝脏炎症坏死并发展为LF。实验研究中主要以动物血清或者人血清白蛋白作为抗原，将其注射入动物体内，复制肝纤维化动物模型。主要造模方法包括以下几种。

（1）动物血清诱发的肝纤维化模型　实验室一般常选用猪血清、牛血清、马血清或血吸虫血清等给予试验动物注射，引起胶原纤维堆积，导致肝小叶消失，形成假小叶，诱发LF。由于猪血清较牛血清价格便宜，而血吸虫血清不易提取、价格昂贵，因此猪血清是实验室肝纤维化造模的首选试剂。有学者采用猪血清（0.5mL/次，2次/周，共12周）对大鼠进行腹部注射，在造模期间每周处死1只大鼠，观察胶原纤维、LF及TGF-β1指标，用来判断大鼠肝纤维化模型成模情况。结果显示，猪血清可成功诱导肝纤维化模型[16-17]。赵宏伟等在传统猪血清诱导LF（腹腔注射，0.5mL/次，共12周）的基础上进行改良，加用高脂高胆固醇饲料（80%玉米粉+20%猪大油+0.5%胆固醇）饲养大鼠，结果发现，实验第4周时大鼠出现LF的现象，提示此法大大缩短了造模时间，方法简便，价格低廉，模型相对稳定[18]。

（2）人血清白蛋白诱发的肝纤维化模型 人血清白蛋白所致的LF分为致敏和攻击两个阶段。谢玉梅等对人血清白蛋白免疫攻击法制备大鼠肝纤维化模型进行改良，使用生理盐水稀释人血清白蛋白，通过等量的不完全福氏佐剂乳化，于尾静脉注射0.5mL/次（含人血清白蛋白4mg），共注射4次，前2次间隔14天，后2次间隔10天，并以30%乙醇作为唯一饮用水；进一步研究发现，在攻击阶段由尾静脉注射改为腹腔注射，或者在注射人血清白蛋白的基础上加用CCl_4，可增加造模成功率[19]。

（3）刀豆素蛋白诱发的肝纤维化模型 刀豆素蛋白诱导LF的机制是刺激T细胞有丝分裂，促进细胞因子（TGF-β、TNF-α等）的释放，引起炎症反应，从而使肝炎发展成LF。有学者将1g/L刀豆素蛋白溶于灭菌的磷酸盐缓冲溶液，按12.5mg/kg体重给予每只大鼠进行尾静脉注射，一周1次，第6次注射1周后处死大鼠，取肝组织分析，发现刀豆素蛋白诱导的大鼠LF与人类肝炎病毒LF接近，随着刀豆素蛋白剂量的增加，肝损伤加重，此法操作简单快捷[20]。

四、胆管结扎造模法

胆管结扎性肝纤维化模型是通过人为结扎肝外胆管使其阻塞，形成LF。其主要机制是由胆管渐进性破坏、胆汁淤积及肝内炎症持续存在，导致肝细胞及胆管细胞进一步损伤，发展为LF。结扎方式为，将大鼠用10%的水合氯醛按照3.5mL/kg体重腹腔注射麻醉，仰卧位固定于手术台上，手术视野消毒、去毛，逐层沿腹正中线打开腹腔，在十二指肠起始端找到胆总管并分离，并于胆总管下方穿两根棉线，分别结扎，并于结扎线中间剪断胆管，闭合腹腔[21]。慕永平等采用胆总管结扎法制备胆汁淤积性肝纤维化大鼠模型，发现该模型可复制人类胆管梗阻造成LF的形态，且方法简便，造模周期短，LF稳定可靠，无须与毒性物质接触[22]。但是该方法不仅会因不同个体对胆管扩张程度的耐受能力不同而使肝实质有不同反应，致使纤维化程度不均一，而且还会因胆汁泄漏引起的腹膜炎而致模型死亡率高，一般在30%以上。

第二节
肾纤维化

肾纤维化是由于肾单位损坏，间质中成纤维细胞大量增生及MF形成、ECM生成并沉积，进而发生肾小球硬化、肾间质纤维化（renal interstitial fibrosis，RIF）的病理过程，最终导致肾功能丧失。确诊肾纤维化的金标准为肾组织的病理学检查，包括观察肾小管、肾间质和肾小球基底膜的病理形态。肾纤维化的大体病理上主要表现为：肾脏明显硬化、体积减小，表面不平整且呈颗粒样变；镜下观察可见肾间质炎性细胞弥漫性浸润、间质纤维组织增生、纤维化形成；肾小球大部分发生硬化和透明样变，毛细血管遭到破坏，系膜基质增生；由于肾小球血流受阻，肾小管发生萎缩；病变较轻处，残余肾代偿性增大。生化检测方面，主要表现在血清尿素、血清肌酐（serum creatinine，Scr）水平及24h尿蛋白定量异常。

一、化学造模法

1. 氯化汞诱导肾纤维化

重金属汞是主要以金属汞、无机汞、有机汞三种形式广泛存在于自然环境中的污染物，对环境和人体有极大的伤害。肾是受无机汞毒性影响的主要器官之一，这是因为无机汞更容易在肾中蓄积，使肾成为受损最严重的器官。

Zalups 等发现，将氯化汞（mercuric chloride，$HgCl_2$）以 $0.5\mu mol/kg$ 或 $2.0\mu mol/kg$ 静脉注射 3h 后，超过总剂量 55% 的 Hg^{2+} 都集中在肾近端小管上皮细胞中[23]。据报道，大鼠以 20mg/(kg·d) 的 $HgCl_2$ 灌胃，1 周即可造成明显肾功能损害，表现为肾间质大量炎性细胞浸润、肾小管膨胀变性，免疫组化显示肾小管的上皮细胞及肾间质大量表达 α-SMA。$HgCl_2$ 短时间染毒尽管显示有 RIF，但所用 $HgCl_2$ 剂量较大，模型动物死亡率较高[24]。$HgCl_2$ 造模时多选取大鼠进行实验，原因是小鼠对 $HgCl_2$ 显示出一定的耐药性，难以建模。本方法可通过皮下注射、静脉注射、灌胃等途径给药。造模中 $HgCl_2$ 剂量至关重要，剂量过小难成模，剂量稍大啮齿类动物容易大批死亡，因此，在预实验时，对于大鼠的剂量，应多进行尝试或采取梯度给药的方式。在国内，多采用每日 8mg/kg $HgCl_2$ 水溶液灌胃给药，8 周左右即可建立肾小管间质纤维化模型。

2. 阿霉素诱导肾纤维化

阿霉素（adriamycin，ADM）为临床上用于化疗的含醌的蒽环类药物。经肾细胞生成，ADM 被还原成半醌型自由基，氧化过程中有活性氧生成，在血小板活化因子等多种脂质介质的作用下，可诱导肾小球上皮细胞（又称足细胞）发生脂质过氧化，使肾小球滤过膜受损，通透性增加，进而影响糖蛋白的代谢；肾小球的滤过屏障由于滤过膜被破坏而受影响，从而造成肾损伤。病变前期以微小病变性肾病综合征为主，后期可进展至局灶节段性肾小球硬化。阿霉素肾病模型为国内、国外目前公认的模拟人类微小病变和局灶节段性肾小球硬化肾病综合征较好的动物模型。

蔺建军等以 4mg/kg ADM 尾静脉注射，1 周后以 2mg/kg 再注射，3 周时间成功造成阿霉素大鼠的肾病综合征模型[24]。亦有报道小鼠尾静脉以单次注射 ADM10mg/kg，6 周后发生间质纤维化[25]。ADM 的组织毒性很强，注射时如果有溢出，大鼠的尾部则易发生缺血性坏死、溃烂等，进而引起感染甚至导致尾部脱落，造成模型大鼠死亡。

3. 马兜铃酸诱导肾纤维化

马兜铃酸（aristolochic acid，AA）属于硝基菲类有机酸类化合物，为细辛、马兜铃、广防己、关木通等中草药的主要成分。AA 诱发肾小管上皮细胞及间质成纤维细胞损伤，导致肾小管萎缩、间质纤维化。

常见造模方法为，大鼠以 5mg/(kg·d) AA 灌胃，连续 8 周后，其肾间质出现增生的成纤维细胞和炎性细胞广泛浸润的情况[26]。另有报道指出，大鼠以 10mg/(kg·d) AA 灌胃，4 周后，肾小管上皮细胞发生凋亡，肾小球基底膜增厚，肾间质出现炎性细胞轻微浸润，且胶原纤维增生明显[27]。

4. 环孢素 A 诱导肾纤维化

钙调磷酸酶抑制剂环孢素 A 为强效免疫抑制剂，临床上肝、心脏、肺、肾等器官移植术后应用较多。肾毒性是环孢素 A 最显著的副反应，主要表现为进行性的肾功能降低以及不可逆性肾小管及血管等组织结构损伤，包括肾小管上皮细胞萎缩，上皮细胞肿胀、变性甚至坏死或脱落，肾间质炎性细胞广泛浸润，肾慢性缺血，间质纤维化呈条带状或灶状，肾小球硬化等情况。

造模方法如下：大鼠以 15mg/(kg·d) 环孢素 A 皮下注射，持续 4 周；或以 30mg/(kg·d) 灌胃，持续 4 周。4 周后，大鼠出现肾小管萎缩和间质炎性细胞弥漫性浸润并发生纤维化；肾小球硬化，肾小动脉内膜显著增生，管腔接近闭塞[28]。

5. 腺嘌呤诱导肾纤维化

腺嘌呤属于嘌呤类含氮杂环化合物，主要用在化疗药物造成的白细胞减少症的相关治疗方面。有研究显示，腺嘌呤在黄嘌呤氧化酶的催化下生成 2,8-二羟基腺嘌呤，后者使肾小管发生堵塞，氮质化合物的排出受影响，造成血清尿素氮、肌酐、尿酸显著升高；过饱和的尿酸在血中生成结晶，在肾小管、肾间质及肾小球部位沉积，形成异物使局部发生肉芽肿性炎症，大量肾单位损伤，最终发生 RIF。

常见造模方法为，大鼠以 0.75% 腺嘌呤饲料喂养 4～6 周，大鼠出现肾小管肿大或萎缩甚至有坏死，肾间质炎性细胞弥漫性浸润并出现广泛纤维化[29]。有学者报道，以 250mg/kg 腺嘌呤灌胃，前 2 周每天 1 次，后 2 周以相同浓度采取每 2 天灌胃 1 次。4 周可复制出肾损伤较重的肾间质纤维化的大鼠模型[30]。

6. 血管紧张素Ⅱ诱导肾纤维化

肾是高血压损伤的重要靶器官之一，高血压所致的肾损伤进一步恶化，会发展为肾功能衰竭。肾素-血管紧张素-醛固酮系统激活及肾局部血管紧张素Ⅱ产生增多为肾纤维化形成与发展的始因。血管紧张素Ⅱ（angiotensin Ⅱ，Ang Ⅱ）作为肾素-血管紧张素-醛固酮系统激活的产物，可升高全身血压及肾小球毛细血管内压，并且作用于肾的 AT1、AT2 受体，可调节细胞增殖、肥大、炎性细胞活化以及纤维化，进而导致肾小动脉硬化或肾实质损害。

造模方法为，将 15mg/mL Ang Ⅱ 溶液灌进胶囊渗透压泵，通过手术将泵植入小鼠背部皮下，以 2000ng/（kg·min）泵出溶液，6 周后，小鼠出现肾小球硬化，同时伴有肾小管扩大、刷状缘脱落、蛋白管型形成、肾间质增多、小动脉增粗硬化等现象[31]。

7. 链脲霉素诱导肾纤维化

链脲霉素是一种从链霉菌中提取的抗生素，能特异性破坏胰岛 β 细胞，诱发糖尿病，高血糖介导代谢失常及血流动力学改变，造成糖尿病性肾病。造模方法为以链脲霉素 200mg/（kg·d）给小鼠腹腔注射，连续 8 周[32]。

二、手术造模法

1. 肾组织大部切除术诱导肾纤维化

该方法的作用机制是，残存肾单位高灌注、高压力、高滤过引起蛋白尿，加速肾功能缺失，造成纤维化形成。该模型因造模技术较简单、手术操作时间短、操作器械易获取、模型成功率高等诸多优点而被广泛采用。

制备方法：准备 6～8 周雄性大鼠，将其随机分为假手术组和模型组，麻醉大鼠，手术操作台固定，去除腹部毛发且确保暴露区域足够大，以 75%乙醇对腹部区域进行消毒。模型组，在距左脊肋角 1～2cm 处斜向外方切一小口以暴露左肾，游离肾周围脂肪组织，用丝线结扎上、下端肾组织约 1/3，

1周后再次手术将整个右肾切除，2次手术一共切除5/6肾组织；假手术组大鼠只需划一切口，不需切除肾组织。模型构建过程中，手术操作要求比较高，操作者应熟练整个手术流程。同时，肾是一个易出血的脏器，应重点做好止血准备。在建立模型后的2周、4周、8周、12周分别取肾组织，通过HE染色、Masson三色染色、PAS染色及免疫组织化学方法观察肾组织结构改变，主要观察肾小球是否有局灶或全部硬化、肾小管是否有萎缩或扩张以及RIF的程度等[33-34]。

2. 缺血再灌注损伤诱导肾纤维化

缺血再灌注损伤为临床上肾移植发病的主要病因之一。短时间缺血引起的损伤可逆，但长时间缺血再灌注后会造成肾组织、细胞结构的不可逆性损伤，导致肾组织纤维化的发生。

制备方法：准备6～8周雄性大鼠，随机分为假手术组和模型组，麻醉大鼠后置于操作台固定，剔除腹部毛发，并用75%乙醇进行消毒。打开腹腔后将双侧肾动脉钳夹45min，然后取下钳夹进行再灌注24h后，取血清和肾组织分别进行研究。操作时，严格把握钳夹动脉血管的时间以及灌注时间，此步是该模型成功的关键。通过HE染色、Masson三色染色和PAS染色，可以观察肾组织结构变化和纤维组织含量变化[35]。

3. 单侧输尿管梗阻诱导肾纤维化

该模型诱导的肾纤维化特征明显，死亡率低，较适用于对RIF相关的研究，是一种研究较成熟的理想肾间质纤维化模型。

制备方法：准备6～8周雄性大鼠，腹腔注射麻醉剂后，将其固定于操作台上，用剃毛器去除腹部毛发，喷洒75%乙醇对腹部区域进行消毒，在左侧腹部行一切口，充分暴露输尿管，并游离周围脂肪组织，切除左侧输尿管，并缝合结扎伤口；假手术组大鼠仅打开腹腔暴露输尿管，并游离周围脂肪组织，但不处理输尿管，分别于手术后3天、7天和14天取肾组织，通过HE染色等观察肾组织结构的改变[36]。手术操作时，应注意严格执行无菌操作，保护好肾周围组织，同时需要将肾放于原位，并逐层缝合。

第三节
肺纤维化

肺纤维化是以炎症和 ECM 沉积为特征，呈进展性和致死性的弥漫性肺间质疾病。在弥漫性间质性肺疾病中，特发性肺间质纤维化发病率最高，且预后极差。IPF 是一种原因不明、以弥漫性肺泡炎和肺泡结构紊乱最终导致肺间质纤维化为特征的疾病，病理组织学特征表现为普通型间质性肺炎，具体表现为上皮细胞增生、基底膜裸露、肺泡实变、出现成纤维细胞灶等。肺纤维化按病因可分为原因明确和原因不明两大类，其中病因不明的 IPF 是临床上最常见且最严重的肺纤维化类型，具有高发病率和高死亡率的特点。

一、化学造模法

1. 博来霉素诱导肺纤维化

目前国际上应用最为广泛的就是博来霉素（bleomycin，BLM）诱导肺纤维化动物模型。BLM 是从轮枝链霉菌中提取的一种抗肿瘤药物，我国从放线菌菌株 72 号中分离出同样的抗生素。作为抗肿瘤药物，BLM 临床应用时常见的副作用是引起肺炎及肺纤维化症状，因而是肺纤维化动物模型最常用的诱导剂。

目前最常用的 BLM 诱导肺纤维化动物模型的方法是一次性气管内灌注，剂量通常是 3～5mg/kg。在 BLM 单次给药后，急性炎症反应会持续 8 天，第 9 天炎症会向肺纤维化转换，28 天或 35 天之后出现组织基质沉积，呈现纤维化的改变[37]。气管内灌注又分为直接灌注和间接灌注。直接灌注法是通过气管插管向动物气管内灌注 BLM；间接灌注法是在动物颈部切口，钝性分离，找到气管后向气管内注入 BLM。两种方法在灌注后都需要将小鼠旋转，以便药物在肺部能够分布均匀，使肺部纤维化程度均匀。李伟峰、王聪等使用 Micro Sprayere 雾化器（购自美国 PENN-CENTURY）气管内雾化的方法建立动物模型，并与气管内滴注方法相比较，结果显示，两种方法均能够引起肺纤维化，但气管内滴注组小鼠肺纤维化程度不一，病灶分布不均；而气管内雾化组小鼠肺纤维化病灶分布均匀，更适合用于小鼠肺纤维化模型[38-39]。国外研究者也使用了这种雾化装置向大鼠气管内雾化 BLM，诱导大鼠肺间质纤维化，降低了大鼠感染的风险，提高了造模成功率及肺内纤维化病灶分布的均匀程度[40]。

2. 异硫氰酸荧光素诱导肺纤维化

异硫氰酸荧光素（fluorescein isothiocyanate，FITC）也可用于诱导小鼠肺纤维化。FITC 可以直接作用于气道，作为半抗原与其他的肺组织蛋白结合，通过趋化因子受体 2（CC chemokine receptor 2，CCR2）与 C-C 基序趋化因子配体 12（C-C motif chemokine ligand 12，CCL12）的相互作用，使纤维细胞进入肺，成为一个长时间的刺激因素，进而诱导肺纤维化。FITC 的发病机制是，FITC 引起急性肺损伤，引发肺水肿和炎症反应，随后发生纤维化。FITC 经气管内给药至肺部，约 14～21 天可以形成肺纤维化[41]。如气管内滴入 FITC 0.007mg/g，纤维化反应可持续 6 个月，可用于长期性肺纤维化研究[42]。

3. 百草枯诱导肺纤维化

百草枯（paraquat，PQ）是一种常用的接触性速效除草剂。肺是 PQ 中毒的主要靶器官，中毒后的继发性损伤产生肺纤维化。PQ 中毒涉及多种机制，现普遍认为氧化应激、炎症损伤是最主要的致病机制。

使用 PQ 建立肺纤维化模型，常见报道有一次性口服灌注、腹腔注射等造模方法。高燕等选用昆明小鼠，通过一次性灌胃 PQ 100mg/kg 的方法建立肺纤维化模型，病理结果提示造模组肺组织 3 天时已出现严重炎性反应，21 天时肺组织呈明显的纤维化改变[43]。一次性口服灌胃方法操作简便，对动物损伤小，可用来进行肺纤维化的发病机理和药物疗效研究。杨珊珊等选用一次性腹腔注射 PQ，尽管 C57BL/6J 小鼠组出现了肺纤维化，但程度轻，造模结果并不理想，可能与 PQ 剂量过小有关[44]。祝春青等通过实验对比了一次性腹腔注射各剂量组的造模效果，结果为各染毒组均出现不同程度的肺纤维化改变，且注射剂量越高，肺纤维化病变越严重，早期死亡率也越高[45]。综合各方面考虑，一次性腹腔注射 PQ 40～50mg/kg 可制备较合适的小鼠肺纤维化模型。

二、物理造模法

1. 辐射诱导肺纤维化

放射性肺纤维化模型的作用机制是，辐射通过使 DNA 损伤直接诱导 Ⅰ 型和 Ⅱ 型肺泡上皮细胞死亡，AMs 涌入到受损区域，随后激活单核细胞，产生炎症因子、促纤维化细胞因子（如 TNF-α 和生长转化因子-β 等），参与肺纤维化的发展过程。

用小鼠造模时通常采用的剂量是 10～20 Gy，局部照射（只照射胸腔），给予辐射 24 周，肺纤维化形成。如给予 C57BL/6J 小鼠胸部 16 Gy 辐射，1～8 周，小鼠肺组织炎性细胞浸润逐渐增多，肺泡壁和肺间质增厚，至 24 周，肺泡结构严重紊乱并出现塌陷，形成纤维化[46]。

2. 高氧诱导肺纤维化

长时间暴露于高氧环境中可导致多器官和组织损伤。肺是高浓度氧吸入后直接作用的靶器官，因此最易受损，从而形成高氧性急性肺损伤并进一步发展成渐进性肺纤维化。

目前，高氧诱导的急性肺损伤动物模型常采用的氧浓度为 95%～100%，暴露时间为 72～96h。长时间暴露于 50%～85% 氧浓度环境中可诱导形成渐进性肺纤维化，例如，将 CD-1 小鼠（3～4 周龄）置于 80% 高氧环境中 168h，小鼠出现严重肺纤维化[47]。

3. 无机粉尘诱导肺纤维化

将无机粉尘［如石棉、二氧化硅（silicon dioxide，SiO_2）等］滴入啮齿类动物肺中可导致纤维结节形成，模拟接触石棉和 SiO_2 的职业人群中出现的石棉肺和硅肺疾病过程。石棉和 SiO_2 均可通过吸入途径和气管内滴入途径进入小鼠肺中，引起肺纤维化。

（1）石棉诱导肺纤维化　小鼠石棉肺模型是少数能建立肺纤维化病灶的模型之一，因此有助于了解肺纤维化的病理发展。但石棉能够引起人罕见的间皮瘤发生，操作人员在应用时应该加强防护。利用石棉气管滴注和雾化吸入两种方式可以诱导小鼠肺纤维化。气管滴注模型在第 7 天出现纤维化、第 14 天纤维化成熟，吸入模型在 1 个月左右才出现纤维化[48]。气管内单次滴注石棉纤维模型的特点是，石棉纤维在两肺叶间分布不均，纤维化常出现在肺中央而不是在胸膜下，建模所需时间短；吸入模型则需要专门的设备，并且建模时间较长。

（2）SiO_2 诱导肺纤维化　SiO_2 滴注到小鼠肺部可以产生纤维结节，与一些职业暴露人群中的肺部广泛的结节性纤维化疾病类似。SiO_2 可以通过雾化吸入、气管滴注或者是口喉抽吸等方式诱导动物肺纤维化，雾化吸入模型常用 C3H/HeN、MRL/MpJ 和 NZB 小鼠[49]，气管滴注则常用 C57BL/6J 小鼠。SiO_2 在肺中沉积，形成持续的毒性刺激炎症反应，使纤维化结节在 SiO_2 周围形成。两种模型相比，气管滴注模型更易操作、更高效，吸入模型与人类硅沉着病的病因更接近，但吸入模型建模需要 40～120 天[50]，而气管内给药建模需 14～28 天[51]。

第四节
心肌纤维化

心肌纤维化是缺血、缺氧、负荷过度、炎症、代谢紊乱等损伤性因素造成的心脏 ECM 过量沉积，以心肌间质胶原含量升高、比例失调及排列紊乱为特征，可导致心肌僵硬度增加及不同程度的心脏舒缩功能障碍；同时，它也是心律失常的结构基础，并与病理性心室重构及慢性心功能不全的进行性发展密切相关。这种情况常见于冠心病、扩张型心肌病、肥厚型心肌病、高血压性心脏病等多种心血管疾病，是这些疾病发展至一定阶段的共同病理改变，也是引起终末期心力衰竭的关键因素。

一、化学造模法

1. 阿霉素诱导心肌纤维化

常见造模方法有：无特定病原体（SPF）级8周龄健康雄性SD大鼠，体重235～265g。实验动物被安置在标准实验室条件下，光/暗周期12h，室温为23℃±2℃。提供标准啮齿动物饲料和水。在23℃±2℃下饲养，自由饮食。随机分为对照组和ADM组。ADM组大鼠通过腹腔注射ADM，每天1次，每次剂量2.5mg/kg，连续2周，建立大鼠心力衰竭模型。对照组大鼠注射等体积的0.9%生理盐水[52]。

SPF级Wistar大鼠，适应性饲养1周。随机分为对照组和ADM组。用0.9%氯化钠溶液将ADM溶解配成浓度为1mg/mL的溶液，腹腔注射ADM 2.5mg/kg，对照组大鼠腹腔注射相同剂量的氯化钠溶液，每周1次，共6次，ADM的总共剂量达到15mg/kg，2周后检测正常组和模型组进行心脏超声，检测左室射血分数、左室舒张期末径、左室收缩期末径、左室短轴缩短率，造模组与正常组比较差异有统计学意义（$p<0.05$），即造模成功[53]。

SD大鼠置于恒温环境下适应性饲养1周后，随机分为对照组和ADM组。ADM组每天分6次腹腔注射ADM，剂量为每次3mg/kg，注射前用生理盐水稀释至0.5mg/mL，连续给药2周，总剂量为18mg/kg，对照组和模型组使用同等剂量的生理盐水腹腔注射[54]。

2. 血管紧张素Ⅱ（AngⅡ）诱导心肌纤维化

AngⅡ诱导心肌纤维化可以将6～8周龄雄性C57BL/6小鼠每天腹腔注射AngⅡ（溶于生理盐水）1.5mg/kg或2.5mg/kg，连续4周，构建心肌纤维化模型[55-56]。也可使用皮下药物渗透泵持续滴注血管紧张素Ⅱ 2.5mg/(kg·d)，连续4周[56]。还可采用外科手术方式为大鼠皮下植入Alzet渗透压缓释泵，每天泵入生理盐水或0.8mg/kg AngⅡ，连续4周，以此分别制作对照组及心肌纤维化模型组[57]。

3. 异丙肾上腺素诱导心肌纤维化

SD 大鼠模型组采用皮下注射异丙肾上腺素的方法构建心肌纤维化模型，每天 2mg/kg，连续注射 10 天[58]。也可每天背部皮下注射异丙肾上腺素 2.5mg/kg（首日 5mg/kg），连续 21 天，建立大鼠心肌纤维化模型，对照组皮下注射等量生理盐水[59]。

4. 链脲佐菌素诱导心肌纤维化

高糖环境诱导心肌纤维化模型是先对动物进行糖尿病造模，糖尿病造模成功的动物会随着糖尿病的病程发展，出现糖尿病心肌病（diabetic cardiomyopathy，DCM）等并发症，进而由 DCM 引发心肌纤维化。目前糖尿病动物造模主要有两种方法：单纯注射链脲佐菌素（streptozotocin，STZ）和高脂饲料配合链脲佐菌素诱导糖尿病模型。

将雄性 SPF 级 SD 大鼠分为正常对照组和高脂饮食（high-fat diet，HFD）组，正常对照组继续饲喂普通饲料，HFD 组改为高脂饲料喂养。饲喂 6 周后，HFD 组大鼠过夜空腹 12h，腹腔注射小剂量 STZ 30mg/kg，STZ 溶液现配现用。0.4% STZ 的配制：在避光、低温、干燥环境中快速称取 1g STZ 溶于 250mL 4℃预冷的柠檬酸盐缓冲液（0.1mol/L，pH=4.0）中，充分溶解后立即使用，30min 完成注射。注射后 72h 检测大鼠随机血糖浓度，随机血糖≥16.7mmol/L 为糖尿病大鼠暂时成模标准。剔除血糖不达标大鼠。1 周后复查暂时成模大鼠，随机血糖仍≥16.7mmol/L 为建模成功，剔除不达标大鼠。并在随后的检测中发现，从第 8 周开始，模型组的心肌羟脯氨酸（hydroxyproline，HYP）、晚期糖基化终末产物含量以及心肌胶原容积分数明显高于对照组，心肌纤维化的特征开始出现；在 12 周时上述各项指标较空白对照组的差异继续增大，表现出明显的心肌纤维化[60]。

SPF 级健康成年雄性 SD 大鼠分为两组：正常对照组和 HFD 组，然后分别予以普通饲料和高糖高脂饲料（配方配制：10% 猪油、20% 蔗糖、6% 蛋白粉、3% 蛋黄粉、61% 常规饲料）饲养 4 周，禁食 12h，高糖高脂饮食大鼠单次腹腔注射 STZ 40mg/kg（溶解于柠檬酸钠缓冲液中，pH=4.4），普通饮食大鼠予以同等剂量柠檬酸钠缓冲液腹腔注射。注射后 3 天和 5 天采尾静脉血测血糖，血糖≥16.7mmol/L 的大鼠为 2 型糖尿病（type 2 diabetes mellitus，T2DM）大鼠，同时出现多食、多尿、多饮症状，提示 T2DM 大鼠模型构建成功[61]。

二、手术造模法

1. 冠状动脉前降支建立心肌梗死诱导心肌纤维化

适应性饲养 3 天，术前 12h 禁食，自由饮水。随机将大鼠分为假手术组和模型组。将大鼠称重，用 5% 水合氯醛按 6mL/kg 剂量腹腔注射麻醉。大鼠麻醉后去毛，仰卧位固定，以碘伏消毒左侧胸前及腋窝下皮肤，胸部备皮。将电极分别插入大鼠右上肢（负极）和左下肢（正极）记录Ⅱ导联心电图。沿左侧第四肋间（或心脏搏动最明显处）做一斜行切口约 1.5cm，用止血钳逐层分离皮下组织，少量钝性分离肋间肌。在心脏搏动最明显处用止血钳轻轻穿破胸膜，撑开胸廓（力度适中，切忌撑断肋骨）。在手术显微镜下，暴露心脏，剥开心包，轻压右胸，将心脏轻轻挤压出胸壁外，持 5/0 无创缝合针丝线，于肺动脉圆锥与左心耳之间，平行左心耳下边缘 1~2mm 处进针，深度为 1.0~1.5mm 进行结扎，结扎力度适中，以防将心肌和血管切断。个别大鼠在冠状动脉结扎后会出现短暂心律失常，绝大多数能够自行恢复，快速心律失常发生时可直接在心脏表面滴少许利多卡因；当心率过慢时可直接心脏按压恢复。清理胸腔，挤出胸腔内空气，用 3/0 的缝合线逐层缝合肌肉，关闭胸腔并对切口消毒。手术过程中保持温度。假手术组大鼠只穿针不结扎冠状动脉，其余操作步骤与模型组相同。模型成功判断标准：结扎后肉眼可见心脏表面相应区域由鲜红色变成苍白色，血管供应范围心肌发绀，室壁活动减弱；结扎后心电表现为 ST 段弓背抬高或（和）T 波高耸或与其形成单向曲线，QRS 波电压增高或（且）波幅增宽均为心肌缺血标志[62]。

SPF 级 SD 雄性大鼠随机分为高位结扎组、低位结扎组和假手术组。10% 水合氯醛按照 0.3mL/100g 的比例进行腹腔注射，腹腔用注射器回抽，主要是确证没有刺入肠道中。在已经麻醉大鼠的四肢皮下插入针电极，电极位置参照人的相应部位，心电图机走纸速度 50mm/s，电压 2mV。麻醉后，用橡皮筋将 SD 大鼠四肢固定于手术台上,用皮筋挂住鼠的门齿,将鼠头略向后仰,做肢体导联体表心电图,然后颈部脱毛,颈部和左侧胸部常规消毒,切开颈部皮肤及浅、深筋膜,用止血钳钝性分离各层肌肉,露出气管,在气管第 3、4 肋软骨间隙切一小口,将呼吸机通气管插入气管约 1.5cm,固定通气管。在左侧胸部常规脱毛、消毒,依次切开皮肤及浅、深筋膜,用止血钳钝性分离胸大肌和前锯肌交界处,在靠胸骨缘处用止血钳钝性分离第 3、4 肋间进胸,剪断第 3、4 肋软骨,撑开肋间露出心脏,用两个带橡皮圈的小拉钩拉开手术切口两侧组织,用湿润的棉签小心推开心脏周围的肺组织,扩大手术视野,用镊子提起心包壁层,再用眼科剪小心剪开心包,棉签向上推开胸腺即可清楚暴露左冠状静脉,以左冠状静脉为标志,用 7/0 眼科无创带线缝合针穿过其深部,进针深度为 0.5~1.0mm 并打结。冠脉高位结扎组在大鼠左心耳与肺动脉圆锥间,距主动脉根部约 3mm 结扎冠状动脉,而冠脉低位结扎组在距离大鼠左心耳尖端约 2mm 水平处结扎冠状动脉,假手术组方法同低位结扎组处理一致,但不结扎冠状动脉,并观察结扎部位以下心肌颜色及心电图的动态变化。仔细检查心脏无出血后关胸,用 0 号丝线逐层间断缝合肋间肌、前锯肌和胸大肌,缝合最后一针时先穿针打虚结,通过此间隙插入 5mL 去针头注射器用于胸腔抽吸,适度增大呼吸机通气量,使肺体积增大,排除胸腔内气体,使胸腔恢复负压,最后一针打结并缝合皮肤。待大鼠出现吞咽动作时,通气管与呼吸机断开,观察 5min,如果无呼吸困难,用注射器抽尽气管内血块和分泌物,拔出通气管,用 7/0 眼科无创带线缝合针缝合一针将相邻两个气管软骨环拉拢,使气管切口闭合,用 0 号丝线依次缝合各层肌肉、深浅筋膜、皮肤。术毕将大鼠放入小饲养笼中 30℃条件下苏醒,每只存活的大鼠放入小饲养笼中单独饲养(主要防止术后因为大鼠的伤口或者血腥味相互打斗)共 7 天,术后肌内注射青霉素预防感染及给予姜黄素腹腔注射治疗。4 周后建模成功[63]。

2. 两肾一夹法诱导心肌纤维化

两肾一夹法是通过保留实验动物双侧肾,对单侧肾动脉进行狭窄,减少单侧肾血流量,激活肾素-血管紧张素-醛固酮系统,从而诱导心肌纤维化。传统两肾一夹法是将大鼠麻醉后,于鼠板上固定四肢,用手术刀沿腹部正中逐层切开进入腹腔，显露左肾动脉。在左肾动脉近中点处放置银夹，使其部分狭窄，而右侧肾动脉不做处理。将大鼠随机分为假手术组和模型组。分笼饲养，室温20~26℃，日照12h，相对湿度40%~70%，通风良好，饲喂普通颗粒饲料，自由饮水。

手术前适应性喂养1周，并在术前12h禁食不禁水。所有手术器械高压蒸汽灭菌备用。手术过程为：用10%水合氯醛（3.5mL/kg）麻醉大鼠，浅麻醉后大鼠俯卧位固定于手术台上，于背部左肾处脱毛，暴露皮肤后用5%乙醇碘酒消毒，以75%乙醇脱碘，然后于触及浮肋即T13肋（大约为背部第3腰椎棘突水平）水平处开始，靠脊椎左侧1cm处作平行于背部后正中线约1.5cm长的手术切口。开口后可立即看到左肾，用无齿手术镊小心将其拉出，并用生理盐水纱布包好推向左侧，此时在近肾门处可以明显看到肾静脉和肾动脉，肾动脉一般位于肾静脉的上后方且极有韧性（辨别时可用无齿小弯镊挑起来，此时呈透明色，放开后立即呈现红色）。用无齿小弯镊钝性分离出左肾动脉（长度约0.5cm），然后穿入无菌丝线（规格为1/0），把直径为0.25mm的针灸针与肾动脉血管长轴紧贴平行放置，然后用无菌丝线扎紧肾动脉和针灸针，以狭窄后左肾颜色变为"浅红色"为宜，然后抽出针灸针，剪去多余的线头，以青霉素钠20万单位冲洗预防感染，逐层缝合伤口，创口消毒处理并去除血迹，手术完成。假手术组除不放置针灸针和不用丝线紧扎外，其余手术操作均同上。术后连续3天给青霉素钠10万单位以防感染，并密切观察皮肤切口、精神状态及进食进水情况。

注意事项：为保证麻醉效果，水合氯醛配制时间不超过4h；开口尽量小，便于缝合；以十字缝合，以免术后裂开感染。术后3天，模型组大鼠血压开始持续升高，与空白组比较差异有统计学意义（$p<0.05$）。与传统方法相比，由背部开口进入以丝线结扎左肾主动脉造模方法的安全性和可操作性更佳[64]。

3. 腹主动脉缩窄诱导心肌纤维化

将8周龄健康SPF级雄性SD大鼠，适应性饲养两天，体重220~270g。随机分为对照组和模型组。所有实验动物以标准大鼠饲料饲养，自由摄食、饮水，温度23~25℃，湿度55%~70%。术前8h禁食，自由饮水。保持实验室室温为25～30℃。大鼠用10%水合氯醛（0.35mL/100g）腹腔注射麻醉，仰卧位，固定四肢，腹部手术区备皮、消毒，并铺一次性无菌纱布。于剑突下2～3cm，腹正中切口，沿腹白线逐层切开皮肤及肌肉至剑突，暴露腹腔。用手指在大鼠左侧背部向腹腔轻推，将胃体、脾脏一并挤出，外翻置于生理盐水浸润的纱布上，固定，随后用脱脂棉球将左侧肝脏向上轻推，充分暴露腹膜以及左侧肾脏。视野可见明显的左肾静脉，于肾静脉水平沿腹主动脉上1.5～2cm处可用手指触及明显搏动感，用两只弯镊在此轻缓钝性分离，剥离腹主动脉血管鞘，分离腹主动脉，随之在其下穿过长约8cm的4/0手术缝线（术前置于生理盐水中湿润），留置备用。将磨钝的7号针头与腹主动脉平行紧贴血管放置，用留置的4/0手术缝线将腹主动脉与7号针头一起结扎，结扎后马上可触及结扎处下方动脉脉搏明显减弱、左肾缺血变白，迅速提起两边缝线同时轻柔地抽出针头，剪去过长的缝线。检查大鼠腹腔内无残留纱布、棉球后，将胃体、脾等小心放回腹腔复位，逐层缝合肌肉和皮肤，消毒。空白对照组大鼠按手术操作分离腹主动脉并穿线，但不做缩窄处理，缝合，消毒。全部大鼠于术后肌注50000U/只青霉素抗感染，连续注射3天。术后密切观察并记录大鼠生存情况、体重以及精神状态、呼吸、饮水、饮食、毛色、活动力等临床表现。术后4周模型组大鼠出现收缩/舒张末期左室后壁厚度和收缩期室间隔厚度增加，8周出现左心射血分数降低，提示心功能开始出现异常；术后8周病理染色结果表明，模型组出现心肌细胞肥大，排列无序、稀疏，心肌间质胶原纤维增多，提示开始出现心肌纤维化[65]。

选择健康成年雄性SD大鼠8～10周龄，体重150～180g，随机分为3组，其中腹主动脉近心端结扎组为A组、腹主动脉远心端结扎组为B组、正常对照组为C组。三组大鼠均以标准饲料喂养，自由觅食饮水。术前常规禁食8h，饮水不受限制。将2.5%戊巴比妥钠（35mg/kg）向SD大鼠腹腔注射麻醉成功后，固定大鼠，腹部手术区常规备皮、消毒，在其腹部右侧铺生理盐水浸湿的无菌纱布，于剑突下2～3cm沿腹中线纵行切开长约3cm切口，沿手术切口将其内脏轻柔翻出腹腔并置于无菌生理盐水浸湿的纱布上。逐层钝性分离至腹主动脉鞘，暴露腹主动脉及双侧肾脏上部，分离出腹主动脉、肾静脉及神经。A组选择在距腹主动脉、肾静脉交汇处以上约3cm处腹主动脉旁；B组选择在距腹主动脉、肾静脉交汇处以上约1.5cm处腹主动脉旁，以上两组分别平行放置钝化7号针头进行标记，用4号丝线将针头与腹主动脉标记处一同结扎。确认肾脏出现缺血外观后，迅速将针头拔出。观察肾脏血流恢复情况后将内脏复位并关闭腹腔。C组只挂线标记不结扎，其余操作同另外两组。手术完毕后，常规向各组SD大鼠腹腔内注射40000U/只青霉素，连续注射3天预防感染，所有大鼠可以自由取水和正常觅食。建模8周后备用。

对比后发现，术后8周，模型组的左心室（left ventricle，LV）射血分数和LV短轴缩短率均低于正常组，且腹主动脉近心端结扎组（距腹主动脉、肾静脉交汇处上约3cm）低于远心端结扎组（距腹主动脉、肾静脉交汇处上约1.5cm），模型组的舒张末期LV后壁厚度高于正常组，且近心端结扎组高于远心端结扎组；对比不同组HE和Masson染色发现，模型组均出现心肌细胞肥大，细胞间质胶原增生，且近心端结扎组的程度严重于远心端结扎组，表明不同的结扎位置所引起的心功能异常和心肌纤维化程度不同，8周后近心端结扎组存活率为60%、远心端结扎组存活率为80%，结扎近心端所引起的心功能异常和心肌纤维化更加严重[66]。

第五节
胰腺纤维化

胰腺具有两大功能，分别是消化和代谢。外分泌腺主要产生胰液，其中含有大量酶；内分泌腺主要是4种胰岛细胞，负责分泌胰岛素和胰高血糖素，以维持机体的血糖平衡。常见的胰腺疾病类型有胰腺炎、糖尿病和胰腺癌等。通过构建有效的动物模型研究各种胰腺疾病，逐渐成为临床机制研究和药物开发的热点。胰腺星状细胞（pancreatic stellate cells，PSCs）是胰腺组织的基质细胞，主要位于胰腺小叶间和腺泡周围区，围绕邻近腺细胞基底部，约占胰腺细胞总数的4%~7%。PSCs在胰腺中有两种形态：静息状态和活化状态。活化状态下的PSCs具有很强的增殖和迁移能力，可促使胰腺纤维化，从而引发胰腺疾病。

一、化学造模法

健康 SPF 级 C57BL/6 雄性小鼠，体重约 $20g\pm2g$，6～8 周龄，将其适应性饲养 2 周后，随机分为对照组和雨蛙素组。雨蛙素组给予腹腔雨蛙素注射 $50\mu g/kg$，每周注射 3 天，每天 6 次，每次间隔 1h，连续注射 6 周，构建胰腺纤维化模型。对照组接受同雨蛙素组等剂量、同频率的 0.9％生理盐水。所有小鼠均为第 6 周处理结束后处死[67]。

或将 C57BL/6J 小鼠随机分为对照组和雨蛙素组。雨蛙素组小鼠均采用反复腹腔注射雨蛙素的方法制备慢性胰腺炎小鼠模型，单次注射剂量为 $50\mu g/kg$，溶于生理盐水 $200\mu L$，每天 6 次，每次间隔 1h，每周一、三、五注射，连续注射 6 周[68]。对照组仅注射与雨蛙素组等剂量、同频率的 0.9％生理盐水[69]。

二、外科手术法

选取大鼠，结扎大鼠胰腺和十二指肠交界处的胰管来诱导慢性胰腺炎，从而避免了胆管及其伴随动脉的损伤。模型鼠在导管结扎后第 2 天接受单次雨蛙素给药（$50\mu g/kg$）。术后 3 天结扎的胰腺部分发生严重的坏死性胰腺炎，模型相关的死亡率为 10％～15％，并始终发生在注射雨蛙素的 48h 内，胰腺坏死区域被纤维化和脂肪组织替代，21 天后检测到纤维化范围最大[70]。

第六节
皮肤纤维化

皮肤纤维化中最具有代表性的疾病是病理性瘢痕和系统性硬化症。病理性瘢痕是深层伤口愈合失衡，以成纤维细胞增生及 ECM 过度沉积为主要表现的疾病，可分为增生性瘢痕与瘢痕疙瘩。其中，增生性瘢痕常见于关节附近，表现为不超过原始创面的红色斑块，一般形成周期为 4～12 周，随着时间的推移逐渐变平；然而瘢痕疙瘩通常出现在缺乏毛囊的皮肤区域，如颈部、胸部、肩部、上背部、耳郭和腹部，且皮损易超出原始创面，并累及周边正常皮肤。瘢痕疙瘩形成周期较增生性瘢痕长，可达数年，无明显消退倾向。系统性硬化症是一种以小血管损害和免疫系统紊乱为特征，导致皮肤和内脏器官纤维化的严重疾病。

一、化学造模法

博来霉素诱导皮肤纤维化通过将博来霉素硫酸盐（100μL，500μg/mL 溶于 NaCl 中）皮内注射到 C57BL/6 小鼠剃毛后的背部皮肤上的单个位置，每天一次，持续 3 周，从而诱导皮肤纤维化。小鼠随机分组：①对照组，小鼠皮内注射 100μL 无菌盐水；②BLM 组，小鼠皮内注射 100μL 博来霉素硫酸盐。BLM 工作液配制：用移液枪吸取 BLM 100μL（100mg/mL）原液于 10mL PBS 中混匀，配制成浓度为 1mg/mL 的 BLM 工作液，放置于 4℃ 保存。原液置于冰箱 −20℃ 保存[71]。

BALB/c 小鼠：6～8 周龄，雌性，SPF 级，体重约 20g±2g；生活环境：通风性好，光照交替，温度 22℃±2℃，相对湿度 40%～70%，进食水与食物每日更换，每日清洗笼具。将小鼠随机分为 2 组，分别为对照组、模型组。对照组：用 1mL 注射器在小鼠背部皮下注射无菌 PBS 100μL，前 4 次依次注射于方框的四角，第 5 次注射于方框中心，以此循环 28 天，每周观察小鼠皮下注射部位皮肤变化。模型组：用 1mL 注射器在小鼠背部皮下注射浓度为 1mg/mL 的 BLM 溶液 100μL，连续 28 天，注射方式同上，每周观察小鼠皮下注射部位皮肤变化[72]。

二、物理造模法

在小鼠身上，于正在愈合的伤口增殖阶段施加额外的物理力，可以诱导形成类似人类增生性瘢痕的皮肤纤维化病变[73]。由此产生的增生性瘢痕可持续6个月。采用膨胀螺钉构建生物力学加载装置，可对愈合伤口施加可控的机械应力。从切口后第4天开始，每隔一天将该装置放置在小鼠背部2cm的线性全层切口上，且不接触伤口本身。为了施加张力，膨胀螺钉在第4天分散2mm，此后每隔一天分散4mm。在该模型中，增生性瘢痕形成的特征表现为由于促生存标志物Akt的激活增强，导致细胞凋亡减少，细胞数量增加[74]。

参考文献

[1] Uno M, Kurita S, Misu H, et al. Tranilast, an antifibrogenic agent, ameliorates a dietary rat model of nonalcoholic steatohepatitis[J]. Hepatology, 2008, 48(1): 109-118.

[2] 叶春华, 刘浔阳. 四氯化碳综合法制备大鼠肝硬化模型[J]. 医学临床研究, 2005, 22(5): 619-622.

[3] Guo Y, Wang H, Zhang C. Establishment of rat precision-cut fibrotic liver slice technique and its application in verapamil metabolism[J]. Clin Exp Pharmacol Physiol, 2007, 34(5-6): 406-413.

[4] Chobert M N, Couchie D, Fourcot A, et al. Liver precursor cells increase hepatic fibrosis induced by chronic carbon tetrachloride intoxication in rats[J]. Lab Invest, 2012, 92(1): 135-150.

[5] 高雅, 韦日明, 黄思茂, 等. 基于TGF-β1/Notch信号通路研究杠板归对二甲基亚硝胺诱导大鼠肝纤维化的作用机制[J]. 中国医院药学杂志, 2017, 37(23): 2318-2321.

[6] 王晶晶, 裴天仙, 郭景玥, 等. 丹参滴丸对二甲基亚硝胺诱导大鼠肝纤维化的治疗作用[J]. 现代药物与临床, 2017, 32(4): 572-578.

[7] 李伟伟, 高海丽, 宋新文, 等. 吡咯烷二硫代氨基甲酸酯在二甲基亚硝胺诱导肝纤维化大鼠中的作用及机制[J]. 中华实用诊断与治疗杂志, 2018, 32(1): 5-9.

[8] 秦冬梅, 李静, 陈文, 等. 硫代乙酰胺诱导大鼠肝纤维化模型合并肾脏病变[J]. 石河子大学学报, 2017, 35(1): 24-128.

[9] 薛改, 刘建芳, 闫成, 等. 硫代乙酰胺诱导大鼠持久性肝纤维化模型的制备[J]. 世界华人消化杂志, 2015, 23(12): 1937-1942.

[10] 张帅, 古维立, 黄迪, 等. 四氯化碳、硫代乙酰胺和猪血清诱导大鼠肝纤维化模型的比较[J]. 中国普通外科杂志, 2012, 21(1): 71-76.

[11] Lieber C S, Jones D P, Decarli L M. Effects of prolonged ethanol intake: production of fatty liver despite adequate diets[J]. J Clin Invest, 1965, 44(6): 1009-1021.

[12] Tsukamoto H, Towner S J, Ciofalo L M, et al. Ethanol-induced liver fibrosis in rats fed high fat diet[J]. Hepatology, 1986, 6(5): 814-822.

[13] 王磊,季光,郑培永,等.大鼠酒精性肝纤维化复合模型的建立[J].中西医结合学报,2006,4(3):282-284.

[14] 高斌,常彬霞,徐明江.慢性酒精喂养加急性酒精灌胃的酒精性肝病小鼠模型(NIAAA模型或Gao-Binge模型)[J].传染病信息,2013,26(5):307-311.

[15] 李丰衣,孙劲晖,祝世功,等.实验性酒精性肝纤维化大鼠模型的建立[C].中华中医药学会第十三届内科肝胆病学术会议论文汇编.杭州:[出版者不详],2008:272-275.

[16] 李璨,陆爽,吴君,等.丹防胶囊对免疫性肝纤维化大鼠肝组织IL-33和ST2表达的影响[J].贵州医科大学学报,2019,44(4):435-440.

[17] 谢君,谢晓芳,李梦婷,等.肝苏颗粒对猪血清致免疫性肝纤维化大鼠肝功能和病理损伤的影响[J].中华中医药杂志,2019,34(2):750-754.

[18] 赵宏伟,柏兆方,张振芳,等.氧化苦参碱抗猪血清诱导的免疫性大鼠肝纤维化作用研究[J].成都大学学报,2015,34(4):319-321.

[19] 谢玉梅,聂青和,康文臻,等.中药双甲五灵冲剂对免疫性肝纤维化大鼠抗肝纤维化的分子机制研究[J].胃肠病学和肝病学杂志,2009,18(2):154-158.

[20] 李鸿立,田聆,魏于全,等.刀豆素蛋白A诱导小鼠肝纤维化模型的建立[J].免疫学杂志,2004,20(5):390-392,396.

[21] Tarcin O,Basaranoglu M,Tahan V,et al.Time course of collagen peak in bile duct-ligated rats[J].BMC Gastroenterol,2011,11:45.

[22] 慕永平,张笑,刘平.黄芪总皂苷通过抑制Notch信号通路干预胆汁性肝纤维化的进展[C].第二十四次全国中西医结合肝病学术会议论文汇编.2015:133.

[23] Zalups R K.Early aspects of the intrarenal distribution of mercury after the intravenous administration of mercuric chloride[J].Toxicology,1993,79(3):215-228.

[24] 蔺建军,杨勇,高娜,等.阿霉素注射次数及剂量对肾病综合征模型的影响[J].中国中西医结合肾病杂志,2011,12(8):676-678,755.

[25] 刘娜,王艳秋,杜丰,等.阿霉素诱导小鼠局灶节段性肾小球硬化肾病模型的研究[J].实用药物与临床,2014,17(12):1536-1540.

[26] 何立群,黄迪,王云满,等.丹酚酸B改善马兜铃酸肾病作用机制的研究[J].西安交通大学学报,2010,31(6):766-769.

[27] Tsutsumi T,Yamakawa S,Ishihara A,et al.Reduced kidney levels of lysophosphatidic acids in rats after chronic administration of aristolochic acid:Its possible protective role in renal fibrosis[J].Toxicol Rep,2015,2:121-129.

[28] 王霞,孙东云,王香婷,等.柴苓汤对慢性环孢素A肾损伤的防护作用及其机制研究[J].中国中西医结合杂志,2012,32(8):1083-1087.

[29] Chang X Y,Cui L,Wang X Z,et al.Quercetin attenuates vascular calcification through suppressed oxidative stress in adenine-induced chronic renal failure rats[J].Biomed Res Int,

2017，2017：5716204.

[30] 陈俊蓉，陈利国，谢林林. 关于腺嘌呤慢性肾衰实验模型的思考[J]. 实验动物科学，2013，30(2)：65-67.

[31] Skibba M, Qian Y, Bao Y, et al. New EGFR inhibitor, 453, prevents renal fibrosis in angiotensin Ⅱ-stimulated mice[J]. Eur J Pharmacol, 2016, 789：421-430.

[32] Ma J, Wu H L, Zhao C Y, et al. Requirement for TLR2 in the development of albuminuria, inflammation and fibrosis in experimental diabetic nephropathy[J]. Int J Clin Exp Pathol, 2014, 7(2)：481-495.

[33] 马园园，刘成海，陶艳艳. 肾纤维化动物模型特点与研究进展[J]. 中国实验动物学报，2018，26(3)：398-403.

[34] Tan R Z, Zhong X, Li J C, et al. An optimized 5/6 nephrectomy mouse model based on unilateral kidney ligation and its application in renal fibrosis research[J]. Ren Fail, 2019, 41(1)：555-566.

[35] 王建军，刘亚楠，王建辉，等. 二乙酰胺三氮脒对肢体缺血再灌注模型小鼠肾损伤的保护作用及其机制[J]. 吉林大学学报，2020，46(1)：14-19，205.

[36] 李丽燕，陈建欧，郑旭旭. 单侧输尿管梗阻肾纤维化大鼠模型中肾组织HGF、MBP-7、TGF-β、CTGF的表达变化[J]. 重庆医学，2020，49(7)：1072-1077.

[37] Moeller A, Ask K, Warburton D, et al. The bleomycin animal model：a useful tool to investigate treatment options for idiopathic pulmonary fibrosis？[J]. Int J Biochem Cell Biol, 2008, 40(3)：362-382.

[38] 李伟峰，胡玉洁，袁伟锋，等. 气管内滴入与雾化博莱霉素致小鼠肺纤维化模型的比较研究[J]. 南方医科大学学报，2012，32(2)：221-225.

[39] 王聪，朱绘明，钱卫平，等. 气道喷雾博来霉素建立特发性肺纤维化模型的研究[J]. 实用老年医学，2013，9(9)：751-755.

[40] Robbe A, Tassin A, Carpentier J, et al. Intratracheal bleomycin aerosolization：The best route of administration for a scalable and homogeneous pulmonary fibrosis rat model？[J]. Biomed Res Int, 2015, 2015：198418.

[41] Moore B B, Paine R 3rd, Christensen P J, et al. Protection from pulmonary fibrosis in the absence of CCR2 signaling[J]. J Immunol, 2001, 167(8)：4368-4377.

[42] Christensen P J, Goodman R E, Pastoriza L, et al. Induction of lung fibrosis in the mouse by intratracheal instillation of fluorescein isothiocyanate is not T-cell-dependent[J]. Am J Pathol, 1999, 155(5)：1773-1779.

[43] 高燕，雍政，黄英，等. 百草枯建立肺纤维化动物模型的实验研究[J]. 空军总医院学报，2003，4：18-19.

[44] 杨珊珊，贾晓民，赵杰，等. 三个品系小鼠百草枯肺纤维化模型的比较[J]. 山西医科大学学报，2014，45(6)：456-459，547.

[45] 祝春青,陈冬波,王喆,等. 腹腔注射百草枯构建小鼠肺纤维化模型[J]. 生物技术通讯,2012, 23(4):563-566.

[46] Liu H, Xue J X, Li X, et al. Quercetin liposomes protect against radiation-induced pulmonary injury in a murine model[J]. Oncol Lett, 2013, 6(2):453-459.

[47] Chen H L, Yen C C, Wang S M, et al. Aerosolized bovine lactoferrin reduces lung injury and fibrosis in mice exposed to hyperoxia[J]. Biometals, 2014, 27(5):1057-1068.

[48] B Moore B, Lawson W E, Oury T D, et al. Animal models of fibrotic lung disease[J]. Am J Respir Cell Mol Biol, 2013, 49(2):167-179.

[49] Davis G S, Leslie K O, Hemenway D R. Silicosis in mice: Effects of dose, time, and genetic strain[J]. J Environ Pathol Toxicol Oncol, 1998, 17(2):81-97.

[50] 周欢. 气管内喷雾注入博莱霉素诱导大鼠肺纤维化模型的建立及其评价[D]. 重庆理工大学,2015.

[51] Lakatos H F, Burgess H A, Thatcher T H, et al. Oropharyngeal aspiration of a silica suspension produces a superior model of silicosis in the mouse when compared to intratracheal instillation[J]. Exp Lung Res, 2006, 32(5):181-199.

[52] 艾文伟,张明,周淑兰,等. 丹酚酸A对阿霉素诱导的心力衰竭大鼠心肌纤维化影响及机制研究[J]. 中国医学创新,2022,19(19):5-8.

[53] 吴琼,董艺丹,王佑华,等. 扩心方通过调节TGF-β1/Smad2通路改善扩张型心肌病大鼠心肌纤维化[J]. 世界科学技术-中医药现代化,2022,24(1):243-251.

[54] 聂连桂,刘茂军,陈坚,等. 硫化氢通过上调神经调节蛋白1/ErbB2抑制阿霉素诱导的大鼠心肌纤维化[J]. 中国老年学杂志,2019,39(9):2197-2200.

[55] 赵倩茹,曹梦菲,孙侠,等. Yes相关蛋白在血管紧张素Ⅱ诱导的心肌纤维化中的作用及机制研究[J]. 中华老年心脑血管病杂志,2022,24(3):306-310.

[56] 辛博,陈力,万丽丽,等. 野黄芩苷对血管紧张素Ⅱ诱导小鼠心肌纤维化的影响[J]. 中医药信息,2018,35(4):4-8.

[57] 李博涛,项羽,崔倩卫,等. PPARγ/NF-κB信号通路参与替米沙坦抑制血管紧张素Ⅱ诱导的大鼠心肌纤维化及改善心功能作用[J]. 山西医科大学学报,2019,50(4):383-388.

[58] 范丽,姚亚妮,李瑜. 地尔硫䓬对异丙肾上腺素诱导的大鼠心肌纤维化的保护作用研究[J]. 心肺血管病杂志,2019,38(7):793-798.

[59] 杜琎,石开虎,赵扬,等. 丹参酚酸B通过抑制PI3K/AKT/mTOR通路促进自噬减轻大鼠心肌纤维化的研究[J]. 现代生物医学进展,2019,19(20):3812-3817.

[60] 侯改霞,肖国强,习雪峰,等. 糖尿病大鼠心肌纤维化与心肌AGEs/RAGE水平的变化[J]. 中国实验动物学报,2018,26(4):461-466.

[61] 曾奇虎,翁静飞,李小林,等. 外源性硫化氢对2型糖尿病大鼠心肌纤维化及TGF-β1/Smads信号通路的影响[J]. 中国免疫学杂志,2020,36(6):653-657.

［62］胡珍，陈景瑞，魏静，等．冠状动脉结扎制备大鼠心肌梗死模型及评价实验研究［J］．天津中医药大学学报，2016，33(2)：90-95．

［63］王勇，高大中，殷跃辉，等．大鼠心肌梗死模型建立方法选择及心电图表现［J］．中国实验动物学报，2011，19(6)：525-529，564．

［64］王文靖，潘毅，杨涛．两肾一夹型高血压大鼠模型的改良及评价［J］．中国实验方剂学杂志，2012，18(1)：203-205．

［65］朱林强，周永焯，谢晓明，等．大鼠腹主动脉缩窄术致大鼠慢性心衰的建模与体会［J］．现代医院，2015，15(10)：21-24．

［66］王博群，何燕，黄慧娟，等．腹主动脉缩窄术不同结扎位置致慢性心力衰竭大鼠模型比较［J］．广西医科大学学报，2019，36(2)：174-178．

［67］谭鹏，陈浩，王安康，等．雷公藤甲素对雨蛙素诱导的慢性胰腺炎小鼠模型胰腺纤维化的影响［J］．临床肝胆病杂志，2020，36(3)：641-645．

［68］Zhang G X, Wang M X, Nie W, et al. P2X7R blockade prevents NLRP3 inflammasome activation and pancreatic fibrosis in a mouse model of chronic pancreatitis［J］．Pancreas，2017，46(10)：1327-1335．

［69］崔立华，张一，白景瑞，等．雨蛙素腹腔注射致小鼠慢性胰腺纤维化病理组织学观察［J］．中国中西医结合外科杂志，2013，19(5)：518-521．

［70］Swain S M, Romac J M, Vigna S R, et al. Piezo1-mediated stellate cell activation causes pressure-induced pancreatic fibrosis in mice［J］．JCI Insight，2022，7(8)：e158288．

［71］Li X H, Zhai Y Q, Xi B R, et al. Pinocembrin ameliorates skin fibrosis via inhibiting TGF-β1 signaling pathway［J］．Biomolecules，2021，11(8)：1240．

［72］夏强．Wnt/β-catenin信号通路在博来霉素诱导的系统性硬化症小鼠血管病变中的作用［J］．中国生物化学与分子生物学报，2022，38(7)：911-918．

［73］Aarabi S, Bhatt K A, Shi Y, et al. Mechanical load initiates hypertrophic scar formation through decreased cellular apoptosis［J］．FASEB J，2007，21(12)：3250-3261．

［74］Paterno J, Vial I N, Wong V W, et al. Akt-mediated mechanotransduction in murine fibroblasts during hypertrophic scar formation［J］．Wound Repair Regen，2011，19(1)：49-58．

CHAPTER 03

第三章
热点植物提取物防治器官纤维化的作用及机制

第一节　槲皮素
第二节　水飞蓟素
第三节　山柰酚
第四节　芹菜素
第五节　淫羊藿苷
第六节　黄芪甲苷
第七节　雷公藤红素
第八节　雷公藤内酯
第九节　青蒿琥酯
第十节　人参皂苷
第十一节　丹参酮ⅡA
第十二节　红景天苷
第十三节　姜黄素
第十四节　白藜芦醇
第十五节　木犀草素

第一节
槲皮素

槲皮素（quercetin，图 3-1）是一种在许多水果、蔬菜和树叶中发现的天然黄酮类化合物。

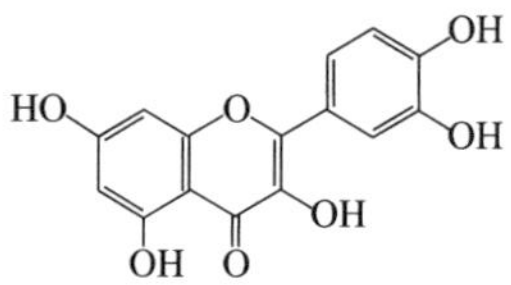

图 3-1 槲皮素分子结构式

1. 肝

槲皮素治疗可以通过信号转导及转录激活因子 3（signal transduction and transcriptional activator 3，STAT3）/细胞因子信号传送阻抑物 3（suppressor of cytokine signaling 3，SOCS3）/胰岛素受体底物 1（insulin receptor substrate 1，IRS1）信号通路减轻 BDL 诱导的 LF 和胰岛素抵抗。此外，槲皮素的抗纤维化和抗炎作用与其调节 RAS 相关 C3 肉毒杆菌毒素底物 1（Ras-related C3 botulinum toxin substrate 1，RAC1）/NADPH 氧化酶 1（NADPH oxidase 1，NOX1）复合物和胞外信号调节激酶 1（extracellular signal-regulated kinase 1，ERK1）/低氧诱导因子-1α（hypoxia inducible factor-1α，HIF-1α）信号转导的能力相关，并且槲皮素还能够抑制 LF 途径中炎症介质的上调[1]。

2. 肺

在原代 IPF 患者的成纤维细胞和正常原代人成纤维细胞中，槲皮素调节小窝蛋白 1（caveolin-1，Cav-1）和凋亡相关因子（factor associated suicide，Fas）表达，并调节蛋白激酶 B（protein kinase B，PKB）活化，使衰老成纤维细胞对凋亡信号的易感性增加。同时，在衰老小鼠的 BLM 诱导肺纤维化模型中，槲皮素抑制肺纤维化的进展，降低衰老细胞标志物和衰老相关分泌表型的表达[2]。在具有肺泡Ⅱ型细胞样表型的 RLE/Abca3 细胞中，槲皮素抑制了 BLM 诱导的 EMT 相关变化和细胞内活性氧水平升高[3]。

3. 肾

大鼠近端肾小管上皮细胞系（renal tubular epithelial cells，NRK-52E）和单侧输尿管结扎诱导肾纤维化模型中，槲皮素激活沉默信息调节因子 2 相关酶 1（silent information regulator factor 2 related enzyme 1，Sirt1），随后在肾小管上皮细胞中诱导 PTEN 诱导激酶 1、人帕金森蛋白 2（human Parkinson disease protein 2，Parkin）介导的线粒体自噬，消除衰老细胞和/或刺激线粒体自噬途径进而预防肾纤维化[4]。槲皮素也可以通过抑制 Hedgehog 信号在体外和体内减轻肾纤维化和 EMT[5]。在单侧肾切除并注射 ADM 的肾小球硬化模型中，槲皮素显著改善了肾小球硬化大鼠的生理指标并改变了 TGF-β 信号通路相关蛋白的表达水平[6]。预注射槲皮素和氯化锂增强了卡维地洛对肾缺血再灌注损伤的肾脏保护作用，而不受脂质介质如磷脂酰肌醇-4,5-二磷酸（phosphatidylinositol 4,5-bisphosphate，PIP_2）和二酰甘油（diacylglycerol，DAG）的影响[7]。

4. 气管

在兔外伤性喉气管狭窄模型和脂多糖（lipopolysaccharide，LPS）诱导的人胚胎肺成纤维细胞（wI-38）中，槲皮素预处理显著逆转 LPS 诱导的促纤维化因子［包括 VEGFs、IL-6、IL-8、胶原蛋白Ⅰ（collagen 1，COL-1）、COL-3、微管相关蛋白 1A/1B 轻链 3A（microtubule-associated protein 1A/1B light chain 3A，LC3）］和纤维化信号介质［如哺乳动物雷帕霉素靶蛋白（mammalian target of rapamycin，mTOR）和 Akt］的上调，并诱导自噬相关基因 5（autophagy-related gene 5，ATG5）的下调。总之，槲皮素通过抑制 TGF-β/Akt/mTOR 信号通路减轻纤维化，进而减轻气管狭窄[8]。

5. 子宫内膜

槲皮素上调微小 RNA-145（microRNA-145，miR-145）并抑制 TGF-β1/Smad2/Smad3 通路的激活，从而调节 TGF-β1 诱导的子宫内膜基质细胞的纤维化反应[9]。

6. 新剂型

槲皮素被配制成水包油 F127 微乳液，提高了其生物利用度和抗氧化活性，减轻庆大霉素诱导的大鼠肾损伤[10]。槲皮素封装到乙型肝炎核心蛋白纳米笼中，作为多功能整合素靶向纳米颗粒可以选择性地将槲皮素传递到活化的 HSCs，用于 LF 的成像和靶向治疗[11]。

第二节
水飞蓟素

1. 心肌纤维化

水飞蓟素（silymarin，图 3-2）通过抑制 TGF-β1/Smad 信号转导来改善 DCM，表明水飞蓟素可能是 DCM 治疗的潜在靶向药物[12]。

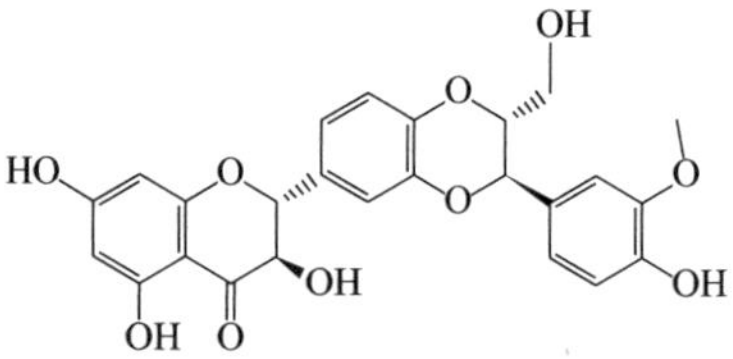

图 3-2 水飞蓟素分子结构式

2. 肝纤维化

水飞蓟素作为一种保肝剂，可调节人肝星状细胞系（human hepatic stellate cell line，LX-2）的促纤维化过程，其效果取决于细胞环境中的压力水平[13]。

3. 肺纤维化

水飞蓟宾是水飞蓟素的主要成分之一，具有显著减少炎症信号、胶原沉积和上皮间充质转分化的作用，表明它可以作为肺部炎症和纤维化的潜在治疗候选药物[14]。

4. 肾纤维化

水飞蓟宾可以在体外和体内显著增加 N-2-(1-羧基-3-苯丙基)-L-赖氨酰-L-脯氨酸（lisinopril，MK-521）的抗纤维化作用。因此，水飞蓟宾与 MK-521 的组合可作为治疗肾纤维化的潜在策略[15]。水飞蓟宾通过抑制核因子-κB（NF-κB）在体外和体内减轻了糖尿病肾病（diabetic nephropathy，DN）的肾纤维化[16]。水飞蓟宾通过抑制 TGF-β1 信号通路显著增加缬沙坦在 TGF-β1 处理的人近端肾小管上皮细胞（human proximal tubular epithelial cells，HK-2）中的抗纤维化作用[17]。

5. 脂肪细胞

水飞蓟宾促进过氧化物酶体增殖物激活受体-α（peroxisome proliferator-activated receptor-α，PPARα）表达以诱导 Sirt1 的激活，从而通过腺苷酸活化蛋白激酶（AMP-activated protein kinase，AMPK）提高小鼠胚胎成纤维细胞（mouse embryonic fibroblasts，NIH/3T3）中的自噬水平。Sirt1 的激活在下调 NF-κB p65 激活中起重要作用，不仅通过使 NF-κB p65 去乙酰化，而且通过减少 NF-κB p65 的磷酸化，导致刺激 AMPK 诱导自噬，进而逆转 COL-1 增强的 3T3-L1 细胞迁移[18]。

6. 新剂型

与传统水飞蓟素片剂相比，油性介质软凝胶胶囊中水飞蓟素-磷脂酰胆碱复合物在健康志愿者中的生物利用度更高[19]。Ⅰ型胶原酶和水飞蓟宾在硫酸软骨素涂层多层纳米颗粒中的共包封用于靶向治疗 LF[20]。壳聚糖包覆的固体脂质纳米颗粒具有优异的稳定性、强大的黏膜黏附性和缓释性，在口服给药后具有增强疏水性水飞蓟宾吸收的潜力[21]。

第三节
山柰酚

山柰酚（kaempferol，图 3-3）是一种黄酮类化合物，化学式为 $C_{15}H_{10}O_6$，其单体纯品为黄色结晶状粉末。

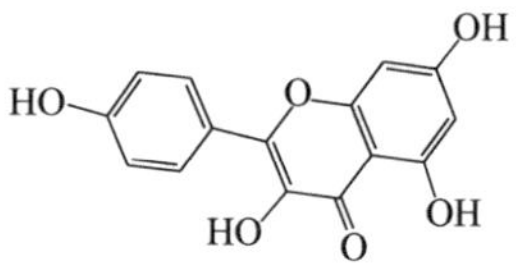

图 3-3 山柰酚分子结构式

1. 心

山柰酚减少了 Ang Ⅱ 诱导的心脏成纤维细胞（cardiac fibroblast，CF）中胶原蛋白的积累，从而抑制了炎症和氧化应激。具体机制为山柰酚靶向 NF-κB/丝裂原活化蛋白激酶（mitogen-activated protein kinase，MAPK）和核转录因子红系 2 相关因子 2（nuclear factor erythroid 2-related factor 2，Nrf2）/AMPK 信号通路治疗心脏重塑和心力衰竭[22]。

2. 肝

山柰酚可以有效地减少 LF 的形成，抑制 HSCs 活化，并进一步抑制体内和体外 HSCs 胶原蛋白的合成。此外，山柰酚可以选择性地结合激活素受体样激酶 5（activin receptor-like kinase 5，ALK5），并进一步下调 TGF-β1/Smads 通路[23]。

3. 肺

山柰酚可部分恢复 SiO_2 诱导的 LC3 脂化而不增加 p62 水平，山柰酚通过调节自噬进而减轻 SiO₂ 诱导的肺纤维化[24]。

4. 肾

在 TGF-β1 处理的 NRK-52E 细胞和单侧输尿管结扎大鼠模型中，山柰酚对 RIF 的保护作用与骨形态发生蛋白-7（bone morphogenetic protein-7, BMP-7）的上调和随后 Smad1/5 的激活密切相关，但与 TGF-β1-Smad2/3 没有直接关系[25]。山柰酚抑制高血糖诱导 Ras 同源蛋白家族成员 A（ras homologous family member A, RhoA）的激活，并减少 NRK-52E 和 RPTEC 细胞中的氧化应激、促炎细胞因子（包括 TNF-α 和 IL-1β 等）和纤维化（TGF-β1、ECM 蛋白水平）的发生。因此，山柰酚可作为 DN 的潜在治疗用药[26]。

5. 皮肤

山柰酚可显著减少 BLM 诱导的皮肤纤维化模型中活性氧（reactive oxygen species, ROS）、MF、T 细胞和巨噬细胞的数量，以及减少炎症和促纤维化细胞因子，包括 IL-6、TGF-β 和 TNF-α。山柰酚还能够通过抑制腺嘌呤核苷三磷酸（adenosine triphosphate, ATP）、细胞外核苷酸的 G 蛋白偶联受体信号，在体内抑制细胞外 ATP 诱导的 IL-6/COL-1 的产生。山柰酚可能是系统性硬化症皮肤纤维化的一种治疗选择[27]。山柰酚的局部应用对糖尿病和非糖尿病大鼠的伤口和切除伤口均有愈合作用；山柰酚通过增加伤口中 HYP 和胶原蛋白的含量、提高伤口抵抗力（即提高抗张强度）、促进伤口闭合和加速再上皮化等方式来发挥这些作用[28]。

第四节
芹菜素

芹菜素（apigenin，图 3-4），也称 4,5,7-三羟基黄酮（4′,5,7-trihydroxyflavone，4′,5,7-THF），是一种黄酮类化合物，存在于蔬菜（欧

芹、芹菜和洋葱)、水果(橙子)、草本植物(洋甘菊、百里香、牛至和罗勒)和植物性饮料(茶、啤酒和葡萄酒)中。

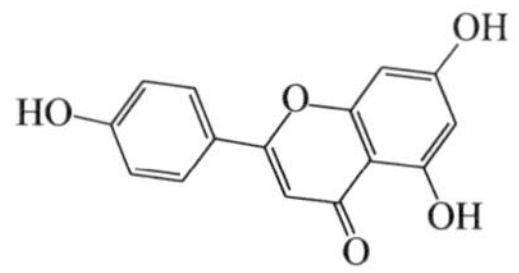

图 3-4　芹菜素分子结构式

1. 心

$4',5,7$-THF 可以抑制 TGF-β1 刺激的 CF 的分化和 ECM 产生,其机制可能部分归因于 miR-155-5p 表达的降低和 c-Ski 表达的增加,这可能抑制 Smad2/3 和 p-Smad2/3 的表达[29]。$4',5,7$-THF 可以抑制 TGF-β1 刺激的 CF 和小鼠心脏纤维化的分化及胶原合成,其机制与 miR-122-5p 表达增加和随后通过直接相互作用下调 HIF-1α 表达有关,这可能最终导致 Smad2/3 和 p-Smad2/3 表达减少、Smad7 表达增加[30]。

2. 肝

在 CCl_4 诱导的 LF 中,$4',5,7$-THF 通过降低天冬氨酸转氨酶(aspartate aminotransferase,AST)、丙氨酸转氨酶(alanine aminotransferase,ALT)、血清碱性磷酸酶(serum alkaline phosphatase,ALP)、乳酸脱氢酶(lactate dehydrogenase,LDH)、HYP、总蛋白(total protein,TP)、总胆红素(total bilirubin,TB)、直接胆红素(direct bilirubin,DB)、透明质酸(hyaluronic acid,HA)、层粘连蛋白(laminin,LN)、血清Ⅲ型前胶原(type Ⅲ procollagen,PC Ⅲ)和Ⅳ型胶原(type Ⅳ collagen,Ⅳ-C)的水平来改善 LF。从机制上讲,$4',5,7$-THF 提高了白蛋白(albumin,ALB)、超氧化物歧化酶(superoxide dismutase,SOD)和谷胱甘肽过氧化物酶(glutathione peroxidase,GSH-Px)的活性,同时降低了丙二醛(malondialdehyde,MDA)的水平。KEGG 通路分析显示 26 个通路包含缺氧诱导因子-1(hypoxia-inducible factor-1,HIF-1)/MAPK/内皮型一氧化氮合酶(endothelial nitric oxide synthetase,eNOS)/VEGF/磷酸化磷脂酰肌醇-3-激

酶（phosphatidylinositol 3-kinase，PI3K）/Akt 信号通路，主要调节肌动蛋白细胞骨架和黏着斑。4′,5,7-THF 可以通过 MAPKs、PI3K/Akt、HIF-1、ROS 和 eNOS 途径，经 VEGF 介导的黏着斑激酶（focal adhesion kinase，FAK）磷酸化改善 CCl_4 诱导的 LF[31]。在体外，4′,5,7-THF 保护 TFK-1 细胞（人胆管癌细胞系）免受过氧化氢（hydrogen peroxide，H_2O_2）诱导的 ROS 产生，并抑制 LX2 细胞（人肝星状细胞系）中的 TGF-β 激活的 COL-1α1 和 α-SMA。在由 3,5-二乙氧基羰基-1,4-二氢-2,4,6-三甲基吡啶（3,5-diethoxycarboxyl-1,4-dihydro-2,4,6-trimethylpyridine，DDC）诱导的胆汁淤积小鼠模型中，4′,5,7-THF 可有效抑制胆囊萎缩、纤维化和胶原蛋白积累。此外，4′,5,7-THF 通过调节法尼醇 X 受体信号通路，缓解 DDC 引起的胆汁酸代谢异常，恢复胆汁分泌与排泄的平衡。此外，4′,5,7-THF 通过阻断 TLR4、NF-κB 和 TNF-α，激活 SOD1/2、GSH-Px 来减少肝脏中的炎症或氧化应激。总之，4′,5,7-THF 通过减少炎症和氧化损伤以及改善胆汁酸代谢来减轻 DDC 诱导的胆汁淤积[32]。

3. 肾

CD38 是哺乳动物组织中主要的 NAD^+ 降解酶之一，4′,5,7-THF 通过抑制 CD38，恢复糖尿病大鼠肾小管细胞内 NAD^+/NADH 比值和 Sirt3 活性，以此改善线粒体氧化应激，从而改善糖尿病引起的肾损伤，例如肾小管间质纤维化、肾小管损伤、炎症和尿白蛋白/肝脏型脂肪酸结合蛋白（liver-type fatty acid-binding protein，L-FABP）排泄[33]。在肾成纤维细胞中，4′,5,7-THF 通过 AMPK 激活和降低 ERK1/2 磷酸化抑制肾成纤维细胞增殖、分化，从而影响其功能，这表明 4′,5,7-THF 可能是一种有潜力治疗肾纤维化的药物[34]。

4. 硬膜

4′,5,7-THF 以浓度依赖性方式抑制成纤维细胞的活力和增殖率，抑制增殖细胞核抗原（proliferating cell nuclear antigen，PCNA）、细胞周期蛋白 D1（cell cycle protein D1，cyclinD1）、Wnt3a 及其下游蛋白的表达。4′,5,7-THF 可通过抑制 Wnt3a/β-连环蛋白（beta-catenin，β-catenin）信号通路而

抑制成纤维细胞增殖，减少硬膜外纤维化（epidural fibrosis，EF）。4′,5,7-THF 可作为预防椎板切除术后减少 EF 的新选择[35]。

第五节
淫羊藿苷

淫羊藿苷（icariin，ICA，图 3-5）是一种从传统中草药（淫羊藿属）中分离出来的活性黄酮类化合物，最初是一种用于增强生殖功能和抗衰老的药物。

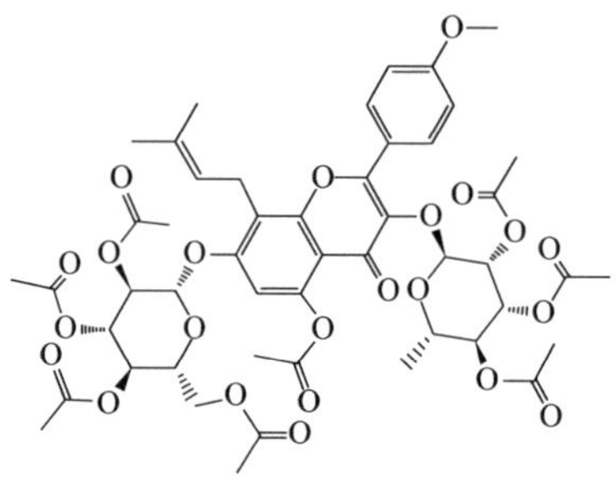

图 3-5 淫羊藿苷分子结构式

1. 心

ICA 减轻了 T2DM 大鼠的心肌结构损伤和纤维化，也抑制了 T2DM 大鼠心肌的细胞内 Ca^{2+} 过度活跃和功能障碍。NOS3、PDE5A 和相关的 sGC-cGMP-PKG 信号通路介导了 ICA 诱导的细胞内 Ca^{2+} 流入改善。此外，ICA 诱导的 c-Jun 和 NF-κB p65 抑制改善了胶原代谢和心肌纤维化[36]。ICA 可预防大鼠异丙肾上腺素（isoproterenol，ISO）诱导的心尖球囊样（takotsubo cardiomyopathy，TCM）综合征心脏功能障碍，其机制主要是通过维持 ROS 系统的动态平衡、促进抗氧化元素活性和抑制 TLR4/NF-κB 信号通路蛋白的表达。此外，ICA 具有的增加抗炎和减少促炎因子分泌的能力，减轻了

TCM 综合征过程中的心肌纤维化[37]。

2. 肝

ICA 对 CCl$_4$ 诱导的 LF 小鼠模型具有保肝作用。ICA 通过直接结合胶质瘤相关癌基因 1 mRNA 的三个主要非翻译区来诱导 miR-875-5p 上调，随后降低胶质瘤相关癌基因 1 的表达[38]。

3. 肾

在单侧输尿管梗阻（unilateral ureteral obstruction，UUO）诱导的慢性肾纤维化小鼠模型中，小鼠在 UUO 手术前连续 3 天和术后连续 14 天口服淫羊藿苷［20mg/(kg·d)］，结果显示，ICA 治疗显著逆转了 UUO 小鼠肾脏中促纤维化因子［如 TGF-β 和结缔组织生长因子（connective tissue growth factor，CTGF）］以及纤维化标志物［如 α-SMA 和纤连蛋白（fibronectin，FN）］蛋白质表达显著升高。ICA 治疗还显著抑制了 UUO 小鼠肾脏中 Smad2/3 的增加和上皮钙黏蛋白（epithelial cadherin，E-cadherin）表达的降低。ICA 治疗显著减少促炎因子［如 NF-κB、环氧合酶 2（cyclooxygenase 2，COX-2）、IL-1β 和 NADPH 氧化酶 4（NADPH oxidase 4，NOX-4）］的蛋白质表达，并增加抗氧化酶［SOD 和过氧化氢酶（catalase，CAT）］的蛋白质表达。ICA 治疗通过其抗纤维化和抗炎特性减轻了慢性肾脏病（chronic kidney disease，CKD）相关的肾纤维化[39]。

第六节
黄芪甲苷

黄芪甲苷（astragaloside Ⅳ，AS-Ⅳ，图 3-6）为环阿尔廷型三萜皂苷类化合物。

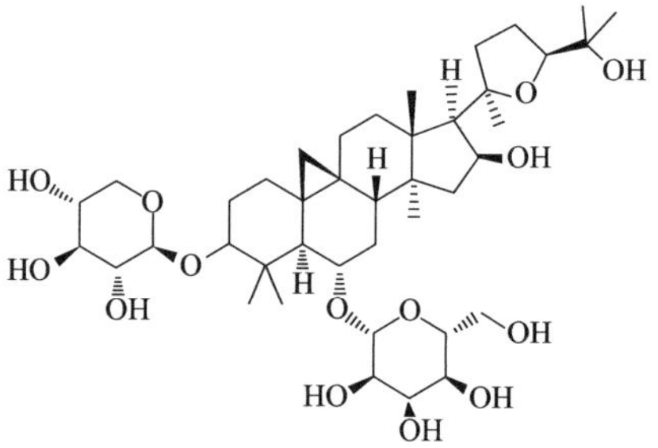

图 3-6 黄芪甲苷分子结构式

1. 心

黄芪提取物具有抑制心脏纤维化的作用，降低 M 型瞬时受体电位通道 7（melastatin-related transient receptor potential 7，TRPM7）的 mRNA 表达。AS-Ⅳ作为黄芪提取物的有效成分之一，在给予相同剂量时表现出相同的效力。AS-Ⅳ通过靶向 miR-135a-TRPM7-TGF-β/Smads 通路抑制心脏纤维化[40]。AS-Ⅳ通过 miR-34a/Bcl-2/（LC3Ⅱ/LC3Ⅰ）和 pAkt/B 淋巴细胞瘤-2（B-cell lymphoma-2，Bcl-2）/（LC3Ⅱ/LC3Ⅰ）通路抑制高糖（high glucose，HG）诱导的氧化应激和自噬，保护心肌细胞免受损伤[41]。与急性心肌梗死大鼠模型组相比，AS-Ⅳ给药显著改善了心脏功能和存活率，减小了梗死面积，减轻了病理变化和纤维沉积，抑制了细胞凋亡，减轻了超微结构损伤和促进血管生成。在体内和体外人脐静脉内皮细胞（human umbilical vein endothelial cell，HUVEC）中，AS-Ⅳ通过调节磷酸酶及张力蛋白同源基因（phosphatase and tensin homolog gene，PTEN）/PI3K/Akt 信号通路发挥心肌梗死后促进血管生成和心脏保护作用[42]。AS-Ⅳ可能对慢性间歇性缺氧（chronic intermittent hypoxia，CIH）引起的心脏功能障碍和结构障碍有治疗作用，这可能归因于调节 Ca^{2+} 稳态和减少心肌细胞凋亡。研究表明，AS-Ⅳ可能是治疗阻塞性睡眠呼吸暂停综合征患者心脏损伤的潜在药物[43]。AS-Ⅳ可抑制体重减轻、心肌损伤、心肌细胞凋亡、心脏纤维化和 ADM 治疗小鼠的心脏功能障碍。AS-Ⅳ通过抑制 NADPH 氧化酶 2（NOX2）和 NOX4 减轻 ADM 诱导的心肌病[44]。AS-Ⅳ减轻 ADM 诱导的大鼠心肌纤维化和心功能不全，抑制大鼠 LVⅠ型和Ⅲ型胶原蛋白、TGF-β、NOX2 和

NOX4 的表达以及 Smad2/3 活性。AS-Ⅳ通过激活 Nrf2 信号通路和谷胱甘肽过氧化物酶-4（glutathione peroxidase-4，GPX4）表达减轻 ADM 诱导的心肌细胞铁死亡。因此，AS-Ⅳ减轻 ADM 诱导的心肌纤维化作用部分归因于其通过增强 Nrf2 信号介导的抗铁死亡作用[45]。

2. 肺

SiO_2 诱导的大鼠肺纤维化模型中，AS-Ⅳ可减轻 SiO_2 诱导的肺纤维化，并降低 COL-1、FN 和 α-SMA 的表达。AS-Ⅳ介导的抗肺纤维化作用可能与炎症和氧化应激的减少有关。机制上，AS-Ⅳ通过 TGF-β1/Smad 信号通路抑制 SiO_2 诱导的肺成纤维细胞纤维化，从而抑制硅沉着病的进展[46]。AS-Ⅳ抑制成纤维细胞中胶原蛋白的表达和向 MF 的转化，具有抗硅沉着病纤维化作用，这可能与抑制 TGF-β1/Smad 信号通路中 Smad3 的持续磷酸化有关[47]。

3. 肾

在永生化小鼠足细胞和糖尿病 KK-Ay 小鼠模型中，AS-Ⅳ通过降低 NF-κB 亚基 p65 乙酰化以及增加 Sirt1 表达来抑制葡萄糖诱导的足细胞 EMT 并增强自噬。在用 AS-Ⅳ治疗后，糖尿病 KK-Ay 小鼠的肾纤维化和肾功能均得到改善。这些发现表明，AS-Ⅳ作为肾脏保护剂，可能通过调节 Sirt1-NF-κB 通路和自噬激活对足细胞 EMT 产生影响[48]。AS-Ⅳ减轻棕榈酸酯诱导的人肾小球系膜细胞（human mesangial cell，HMC）中的脂质积累，抑制 ROS 产生，抑制 TGF-β1、p-Smad2/3、FN、COL 4A1、NOX4 和 p22phox 的表达。AS-Ⅳ通过下调 CD36 表达、介导游离脂肪酸摄取和脂质积累来抑制棕榈酸酯诱导的 HMC 氧化应激和纤维化[49]。AS-Ⅳ治疗改善了 DN 大鼠的临床症状和病理变化，并改善了肾功能。转录组学分析显示，AS-Ⅳ治疗显著减轻了大鼠肾脏中 DN 相关氧化应激和炎症相关核苷酸结合寡聚化结构域样受体（nucleotide-binding oligomerization domain-like receptors，NLR）信号通路的转录。进一步研究表明，AS-Ⅳ治疗减弱了 NLR 信号转导并延缓了 DN 大鼠的肾纤维化进程[50]。AS-Ⅳ治疗降低了 HK-2 细胞的 EMT 特征，抑制了 mTORC1/70 kDa 核糖体蛋白 S6 激酶（70 kDa ribosomal protein S6 kinase，p70S6K）通路的激活、snail 和 twist 的表达下调及 FN 和 COL Ⅳ 的分泌。总之，研究结果表明，AS-Ⅳ 通过阻断 HK-2 细胞中的 mTORC1/p70S6K 信号通路来改善高葡萄糖介导的肾小管 EMT[51]。

4. 腹膜

AS-Ⅳ 可提高细胞活力，抑制高糖腹膜透析（peritoneal dialysis，PD）液中腹膜间皮细胞（peritoneal mesothelial cells，PMCs）的凋亡和 EMT。AS-Ⅳ 通过抑制视黄酸 X 受体 α（retinoic acid X receptor α，RXRα）维持 PMCs 活力以抗细胞凋亡并抑制 EMT，进而发挥抗腹膜纤维化作用[52]。

5. 眼

AS-Ⅳ 的抗纤维化作用是通过抑制 NF-κB 和激活基质金属蛋白酶（matrix metalloproteinase，MMP）来介导的。AS-Ⅳ 治疗还降低了稳定表达突变肌纤蛋白的 TM3 细胞中的内质网（endoplasmic reticulum，ER）应激。局部眼部 AS-Ⅳ 滴眼液（1 mmol/L）显著降低小鼠 TGF-β2 诱导的高眼压，这与小鼠小梁网组织中 FN、COL-1（属于 ECM 成分）、KDEL（ER 应激）和 α-SMA 的降低有关。总之，结果表明，AS-Ⅳ 通过调节小梁网中的 ECM 沉积和 ER 应激来预防 TGF-β2 诱导的高眼压[53]。

第七节
雷公藤红素

雷公藤红素（celastrol，图 3-7）是一种从雷公藤中提取的五环三萜。

1. 心

在大鼠原代心肌细胞和 H9C2 细胞中，雷公藤红素减弱 Ang Ⅱ 诱导的细胞肥大和纤维化反应。雷公藤红素直接结合并抑制 STAT3 磷酸化和核转位。雷公藤红素通过靶向 STAT3 减轻 Ang Ⅱ 诱导的心脏重塑[54]。

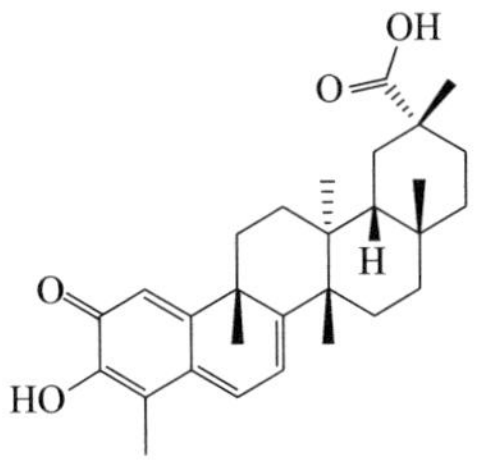

图 3-7　雷公藤红素分子结构式

2. 肝

雷公藤红素在活化的 HSCs 和纤维化肝脏中均显示出对 LF 的有效改善作用。雷公藤红素在体内显著抑制炎症反应，在体外可抑制炎症因子的分泌。同时，雷公藤红素增加了纤维化肝脏和活化的 HSCs 中的 Sirt3 启动子活性、Sirt3 表达和 AMPK 磷酸化。雷公藤红素主要通过激活 AMPK-Sirt3 信号通路抑制炎症来减轻 LF，这使得雷公藤红素成为治疗或预防 LF 的潜在候选化合物[55]。

3. 肺

基质金属蛋白酶诱导因子（basigin，Bsg）是一种跨膜糖蛋白，可促进 MF 分化、细胞增殖和 MMP 活化。亲环素 A（cyclophilin A，CyPA）与其受体 Bsg 结合，促进肺动脉平滑肌细胞增殖和炎症细胞募集。雷公藤红素能够抑制肺动脉平滑肌细胞增殖、ROS 生成和炎性细胞因子释放，在小鼠缺氧诱导的 PH 和 SU5416/缺氧诱导的大鼠 PH 模型中，雷公藤红素可降低右心室收缩压、缓解右心室肥大、减少纤维化并改善右心室功能障碍，降低了心肺中的 Bsg、CyPA 和炎性细胞因子水平。这些结果表明，雷公藤红素通过抑制 Bsg 及其配体 CyPA，减轻肺动脉高压的右心室中因 Bsg 升高而加剧的右心室功能障碍[56]。

4. 肾

使用 FN 结合脂质体将雷公藤红素靶向递送至肾间质肌成纤维细胞，可减轻肾纤维化并降低雷公藤红素的全身毒性[57]。

5. 系统硬化

雷公藤红素是一种新型的大麻素受体 2（cannabinoid receptor 2，CB2）选择性激动剂，在系统性硬化症小鼠模型中具有抗炎和抗纤维化活性。雷公藤红素以剂量依赖性方式触发多个 CB2 介导的下游信号通路，包括环磷酸腺苷（cyclic adenosine monophosphate，cAMP）积累抑制和受体脱敏，并且对大麻素受体 1（cannabinoid receptor 1，CB1）具有中等选择性。此外，雷公藤红素对 LPS 处理的鼠 RAW264.7 巨噬细胞和原代巨噬细胞表现出抗炎特性。因此，雷公藤红素在 BLM 诱导的系统性硬化症小鼠模型中发挥抗纤维化和抗炎作用[58]。

6. 关节

胶原诱导的类风湿关节炎（rheumatoid arthritis，RA）Wistar 大鼠模型中，以雷公藤红素治疗可以抑制炎症细胞因子的释放，包括 TNF-α、IL-6、IL-1β，以及抑制 Bcl-2 相关 X 蛋白（bcl-2-associated X protein，Bax）、Cleaved caspase-3、COL-1、COL-3 和 α-SMA 的表达，减弱心肌细胞的细胞凋亡和纤维化，并提高 Bcl-2 的表达。雷公藤红素可通过调节 TLR2/高迁移率族蛋白 1（high mobility group protein 1，HMGB1）信号通路抑制自噬，从而在类风湿关节炎中发挥心脏保护作用[59]。

7. 皮肤

雷公藤红素给药诱导 NIH/3T3 的生长抑制，表现为 MMPs、VEGF 和碱性成纤维细胞生长因子（basic fibroblast growth factor，bFGF）表达降低的趋势。这个过程伴随着 p65 的磷酸化减少、IκBα 的抑制剂和 β-catenin 的下调。雷公藤红素通过抑制 ANRIL 在 NIH/3T3 中表现出抗瘢痕形成的生物活性，该过程伴随着对 NF-κB 和 β-catenin 级联的阻断[60]。

第八节
雷公藤内酯

雷公藤内酯（triptolide，图 3-8）是从雷公藤中分离得到的二萜内酯化合物。

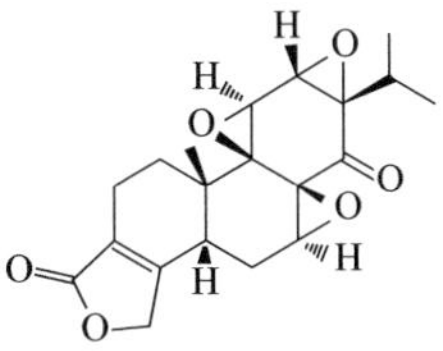

图 3-8 雷公藤内酯分子结构式

1. 心

雷公藤内酯剂量依赖性地抑制异丙肾上腺素诱导的心脏纤维化和血管紧张素 Ⅱ 诱导的 CF 的胶原生成。雷公藤内酯通过抑制 NOD 样受体蛋白 3（NOD-like receptor protein 3，NLRP3）炎性体的活化而发挥抗纤维化作用，具体表现为抑制 IL-1β 成熟、髓样分化蛋白 88（myeloid differentiation protein 88，MyD88）相关的 c-Jun 氨基端蛋白激酶（c-Jun N-terminal kinase，JNK）磷酸化以及 ERK1/2 和 TGF-β1/Smad 信号，并最终导致胶原蛋白生成减少。雷公藤内酯通过抑制 NLRP3-ASC 相互作用来抑制炎性体激活，从而抑制 NLRP3 和含有半胱天冬酶募集结构域凋亡相关斑点样蛋白（apoptosis related spotted protein，ASC）以及炎性体组装的 ASC 的表达。雷公藤内酯在抑制 NLRP3 炎性体的激活以减轻心脏纤维化方面发挥重要作用[61]。雷公藤内酯纳米混悬液均能降低心肌纤维化程度，延缓 LV 重构过程，进一步改善心功能[62]。

2. 肝

在蛋氨酸/胆碱缺乏（methionine-choline deficient，MCD）饮食诱导的

非酒精性脂肪性肝炎（nonalcoholic steatohepatitis，NASH）模型中，每天腹腔注射 50μg/kg 雷公藤内酯可显著减轻肝损伤，但 100μg/kg 雷公藤内酯会引起严重的肝毒性。雷公藤内酯治疗可减少 NASH 中的肝脏脂质沉积、炎症和纤维化，降低转氨酶和胆红素水平。低剂量雷公藤内酯可抑制 MCD 诱导的纤维化（FN、α-SMA、COL、TGF-β）和炎症［ILs、TNF-α、单核细胞趋化蛋白-1（monocyte chemoattractant protein-1，MCP-1）］相关的典型基因和蛋白质的表达。进一步研究表明，雷公藤内酯作为变构 AMPK 激动剂具有改善肝脏脂质代谢、炎症和纤维化，减轻非酒精性脂肪肝（nonalcoholic fatty liver disease，NAFLD）的作用[63]。

3. 肺

雷公藤内酯通过干扰辐射诱导的 IKKβ 激活，阻止了 NF-κB 核易位和 DNA 结合，并有效降低了赖氨酰氧化酶合成，减轻了辐射诱导的肺纤维化[64]。

4. 肾

在醋酸去氧皮质酮（deoxycortone acetate，DOCA）-盐处理的小鼠中，雷公藤内酯治疗以剂量依赖性的方式抑制尿白蛋白和 8-异前列腺素排泄的增加，同时减轻肾小球硬化、肾小管间质损伤和纤维化。此外，雷公藤内酯治疗剂量依赖性地抑制 DOCA-盐诱导的间质单核细胞/巨噬细胞浸润，这与肾脏促炎细胞因子/趋化因子和黏附分子水平以及肾脏激活的 NF-κB 浓度降低相关。雷公藤内酯治疗可能通过减轻盐敏感性高血压的炎症反应来改善肾损伤[65]。

5. 硬膜

雷公藤内酯可以抑制 PI3K/Akt/mTOR 信号通路的激活。同时，雷公藤内酯可以抑制成纤维细胞增殖，诱导成纤维细胞凋亡和自噬，被称为两次细胞自我毁灭。雷公藤内酯可以减少 EF 形成，其潜在机制可能是通过抑制 PI3K/Akt/mTOR 信号通路来抑制成纤维细胞增殖，刺激细胞凋亡和自噬[66]。

6. 回肠

雷公藤内酯通过调节 miR-16-1/热休克蛋白 70（heat shock protein 70，HSP70）通路抑制克罗恩病患者回结肠吻合口成纤维细胞的迁移和增殖[67]。

第九节
青蒿琥酯

青蒿琥酯（artesunate，ART，图 3-9）是从传统中草药艾蒿中提取的青蒿素中提取的小分子物质。

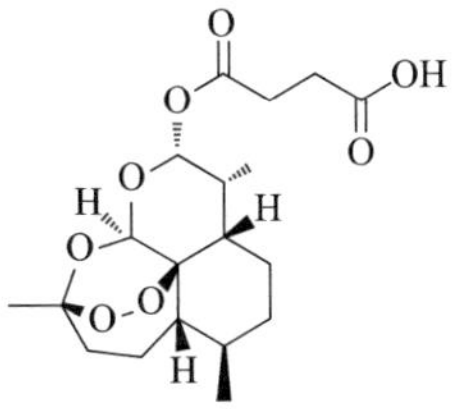

图 3-9 青蒿琥酯分子结构式

1. 肝

在血吸虫病感染小鼠建立的肝纤维化模型中，ART 显著抑制 LF。在 HSCs 细胞系 LX-2 中，ART 有效抑制细胞增殖活性和包括 COL-1a1 和 COL-3a1 在内的纤维化标志基因的 mRNA 表达。ART 以剂量依赖性方式抑制线粒体柠檬酸循环中的酶，例如柠檬酸合酶、异柠檬酸脱氢酶和 α-酮戊二酸脱氢酶。ART 降低 HSCs 线粒体耗氧率和线粒体复合体 I 亚基 NDUFB8 和复合体 III 亚基 UQCRC2 蛋白水平。研究表明，ART 可能通过抑制线粒体中的 NDUFB8 和 UQCRC2 来下调 HSCs 活性，从而减轻 LF[68]。

2. 硬膜

ART 可以诱导自噬并抑制成纤维细胞增殖，这一现象与肿瘤蛋白 53

(tumor protein 53，p53)、细胞周期调控蛋白（cell cycle regulatory protein，p21WAF1）和 cip1 蛋白的上调有关。ART 对大鼠椎板切除术后的 EF 具有显著的抑制作用。总之，ART 可能是通过自噬级联介导的 p53/p21WAF1/Cip1 通路抑制椎板切除术后成纤维细胞增殖，减少 EF 形成[69]。

3. 关节

ART 诱导细胞自噬通量并抑制成纤维细胞中的细胞增殖与 PI3K/Akt/mTOR 通路和 AMPK/mTOR 通路介导的 mTOR 信号转导抑制有关。在体内，ART 治疗触发了自噬激活并减轻了手术引起的膝关节纤维化。总之，ART 通过抑制 mTOR 信号转导诱导 Beclin-1 介导的自噬，在成纤维细胞中表现出抗增殖功效，并减轻了兔膝关节纤维化[70]。局部注射 ART 可以防止关节瘢痕粘连。可能的机制是通过经典的 ERS 途径中内质网应激蛋白激酶 R 样内质网激酶（protein kinase R-like endoplasmic reticulum kinase，PERK）信号通路诱导成纤维细胞凋亡。研究表明，在 ART 治疗后，成纤维细胞中 ERS 被激活，葡萄糖调节蛋白 78（glucose regulated protein 78，GRP78）、增强子结合蛋白同源蛋白和 Bax 表达上调，而 Bcl-2 表达下调。此外，ART 使 PERK 通路中的 p-PERK 与 PERK、p-eIF2α 与真核细胞起始因子 2α（eukaryotic initiation factor 2α，eIF2α）的比例以浓度和时间依赖性上调，表明 PERK 途径参与 ART 介导的细胞凋亡[71]。

4. 皮肤

ART 通过抑制兔耳肥厚性瘢痕模型中的 TGF-β/Smad3 信号通路促进伤口愈合、减少瘢痕组织的形成与增生，并减少 ECM 的产生。ART 以剂量依赖性方式降低了 TGF-β1/Smad3 诱导的成纤维细胞增殖和胶原蛋白生成[72]。在 CRL-2097 人真皮成纤维细胞中，ART 下调促纤维化基因的表达，包括典型的 MF 标志物、ECM 基因和几种 TGF-β 受体和配体，并上调细胞周期抑制剂和 MMP 的表达。总之，这些数据表明 ART 拮抗成纤维细胞活化并降低促纤维化基因的表达[73]。

第十节
人参皂苷

1. 人参皂苷 AD-1

AD-1 是一种源自人参的新型人参苷。AD-1 可与几种选定的氨基酸进行结合,其中结合物 7c 在活化的 t-HSC/Cl-6 细胞中显著抑制细胞增殖,并且对 LO2 无毒。此外,结合物 7c 具有较好的血浆稳定性。进一步的研究表明,结合物 7c 通过诱导 S 期阻滞和激活线粒体介导的细胞凋亡发挥抗纤维化作用[74]。

2. 人参皂苷 AD-2

AD-2(图 3-10)可改善 Ang Ⅱ 诱导的高血流动力学和 P 波异常,并防止纤维化胶原沉积。AD-2 抑制 Ang Ⅱ 诱导的核因子 κB、激活蛋白 1 和 TGF-β1 的表达上调,以及缝隙连接蛋白 40(connexin 40,Cx40)和 Cx43 表达下调,同时通过肝激酶 B1 激活增加 AMPK 的表达[75]。AD-2 可以降低 Ⅰ 型胶原蛋白、基质金属蛋白酶抑制剂-1(tissue inhibitor of matrix metalloproteinase 1,TIMP-1)和 MMP-13 的表达,抑制 ECM 沉积,起到抗 LF 的作用。AD-2 的抗纤维化作用与 LF 相关的炎症因子[包括 TNF-α、IL-1β、半胱氨酸-天冬氨酸蛋白酶-1(cysteinyl aspartate specific proteinase-1,caspase-1)和 IL-6]以及 p-JNK 和 p38-ERK 通路有关[76]。

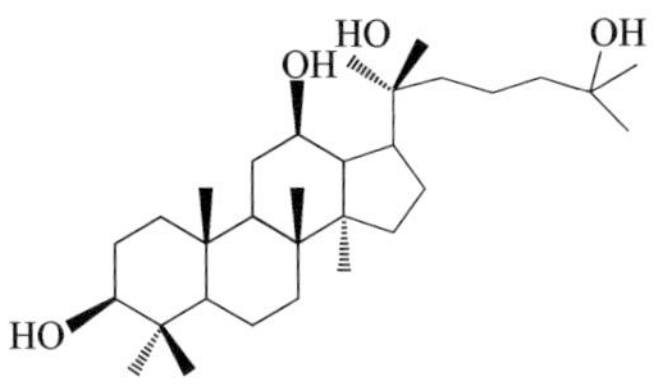

图 3-10 人参皂苷 AD-2 分子结构式

3. 人参皂苷 Rb1

在心肌缺血再灌注（ischemia-reperfusion，IR）模型中，人参皂苷 Rb1（ginsenoside Rb1，G-Rb1，图 3-11）减少了线粒体 ROS 的产生，减少了心肌梗死面积，保留了心脏功能，并限制了心脏纤维化。G-Rb1 对线粒体复合物Ⅰ依赖性呼吸和再灌注诱导的 ROS 产生的抑制作用，可通过使酵母 NADH 脱氢酶绕过复合物Ⅰ得以挽救。分子对接和表面等离子共振实验表明，G-Rb1 可能是通过与 ND3 亚基结合，在再灌注时以失活形式捕获线粒体复合物Ⅰ，降低了 NADH 脱氢酶活性。因此，G-Rb1 通过抑制线粒体复合物Ⅰ介导的 ROS 爆发减轻心脏 IR 损伤[77]。在单侧输尿管结扎模型中，G-Rb1 治疗可显著抑制自噬，并缓解梗阻性肾小管损伤小鼠的纤维化进展。就机制而言，G-Rb1 通过调节 Akt 非依赖性和 AMPK 依赖性 mTOR 信号通路发挥抑制自噬的作用。同时，ERK、p38 MAPK 信号通路也参与了 G-Rb1 对自噬的抑制[78]。

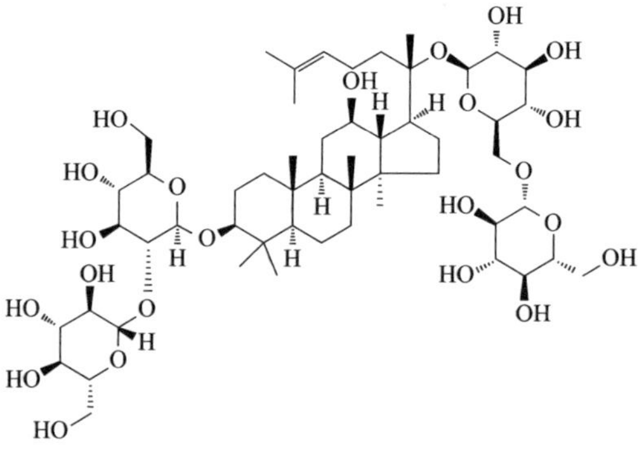

图 3-11　人参皂苷 Rb1 分子结构式

与对照条件相比，G-Rb1 治疗显著增加了老年小鼠收缩末期和舒张末期 LV 的内径和容积，并显著增加了射血分数百分比。G-Rb1 治疗减少了老年小鼠心脏中的胶原蛋白沉积，并且降低了 COLⅠ、COLⅢ和 TGF-β1 蛋白的表达水平，减轻炎症反应，包括降低血清或心肌 IL-6 和 TNF-α 水平。此外，老年小鼠的 G-Rb1 处理增加了细胞质 NF-κB 但减少了核 NF-κB，这表明 G-Rb1 通过调节 NF-κB 的易位来抑制 NF-κB 信号通路。Rb1 可能通过调节

NF-κB信号通路抑制纤维化和炎症来缓解衰老相关的心肌功能障碍[79]。

4. 人参皂苷 Rb3

壳聚糖负载于三聚磷酸钠的纳米粒子与人参皂苷共轭可直接参与心肌能量代谢的重塑和PPARα的调节，从而改善心肌纤维化程度。NpRb3通过靶向PPARα通路改善脂肪酸氧化、糖酵解和调节线粒体氧化磷酸化水平来抑制心肌纤维化[80]。

5. 人参皂苷 Rd

G-Rd（图3-12）显著改善了压力超负荷引起的小鼠收缩功能障碍、纤维化、心脏肥大、炎症和氧化应激。此外，与对照相比，G-Rd处理的心脏中，Akt、钙调神经磷酸酶A、ERK1/2和TGF-β1的蛋白质水平显著降低。G-Rd抑制PE诱导的心肌细胞肥大[81]。

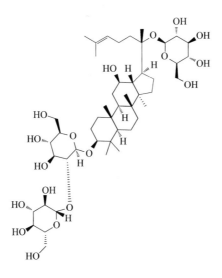

图3-12　人参皂苷Rd分子结构式

6. 人参皂苷 Re

左冠状动脉前降支结扎诱导的心肌梗死（myocardial infarction，MI）

模型中，G-Re（图 3-13）显著抑制 MI 大鼠的心肌损伤并减轻氧化应激，显著改善心功能，防止 LV 扩张，显著降低 LV 间质纤维化。与 MI 组相比，G-Re 治疗促进 AMPKα 磷酸化，降低 TGF-β1 表达，并减弱 Smad2/3 活化，增强 FAK、PI3K p110α 和 Akt 的磷酸化。这些结果表明，G-Re 可能通过调节 AMPK/TGF-β1/Smad2/3 和 FAK/PI3K p110α/Akt 信号通路来改善 MI 引起的心功能不全并减轻心室重构[82]。

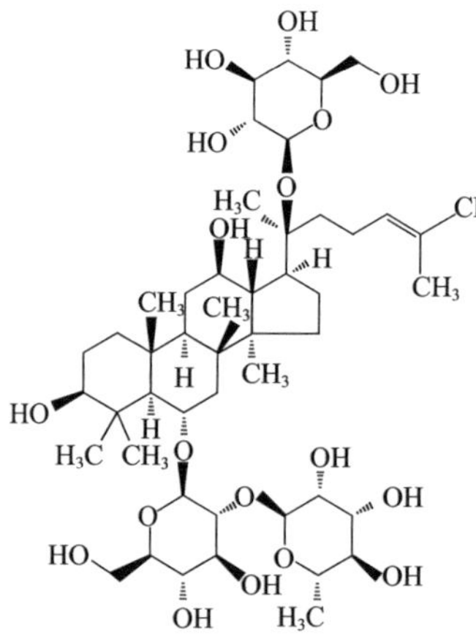

图 3-13　人参皂苷 Re 分子结构式

7. 人参皂苷 Rg1

G-Rg1 可作为吲哚胺-2,3-双加氧酶 1（indoleamine 2, 3-dioxygenase 1，IDO1）抑制剂，改善肝功能和缓解 LF。G-Rg1 减轻 IDO1 介导的 DCs 成熟抑制而发挥抗纤维化特性[83]。

8. 人参皂苷 Rg2

G-Rg2（图 3-14）改善了心肌梗死后小鼠的心脏功能并抑制了胶原蛋白的沉积。此外，G-Rg2 通过激活 Ang Ⅱ 诱导的 CF 中的磷酸化 Akt，降低了纤维化相关基因 COL-1、COL-3 和 α-SMA 的水平[84]。

在异丙肾上腺素诱导的大鼠心肌缺血模型中，G-Rg2 通过调节 TGF-β1/Smad 信号通路，发挥减轻心肌纤维化和改善心功能的作用[85]。

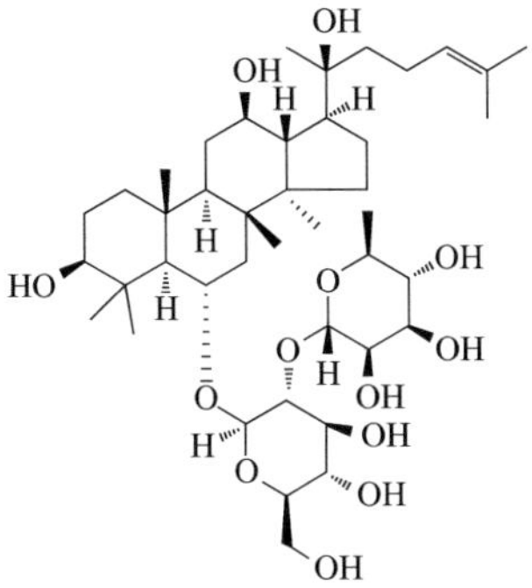

图 3-14 人参皂苷 Rg2 分子结构式

9. 人参皂苷 Rg3

人参皂苷 Rg3（ginsenoside Rg3，G-Rg3，图 3-15）作为原人参二醇皂苷成分之一，在高丽参（*Panax ginseng* CA Meyer）中含量相对丰富。

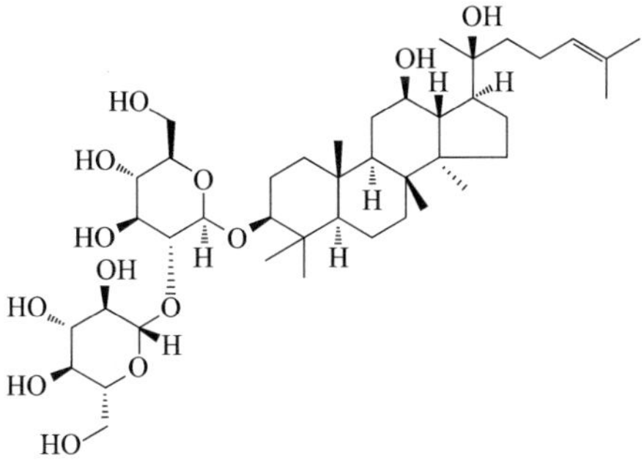

图 3-15 人参皂苷 Rg3 分子结构式

在 TAA 诱导慢性小鼠肝纤维化模型中，G-Rg3 治疗减轻了肝脏病理变化和 LF，减少胶原纤维沉积，减少 HSCs 活化标志物（α-SMA）的表达，并减少促纤维化因子（TGF-β1）的分泌，降低了自噬相关蛋白的表达。G-Rg3 在体外剂量依赖性地抑制自噬。在炎症诱导剂 LPS 诱导的大鼠 HSC-T6 细胞中，p62 的表达减少，LC3a 向 LC3b 的转化也减少。此外，G-Rg3 在体内和体外增强了 PI3K 和 Akt 的磷酸化。综上所述，G-Rg3 通过抑制 TAA

处理的小鼠和 LPS 刺激的 HSC-T6 细胞的自噬发挥抗纤维化作用[86]。为提高药物利用率。使用聚乙二醇和聚硫化丙烯的二嵌段共聚物用于 G-Rg3 封装和递送。在大鼠缺血再灌注模型中，心肌内注射负载 G-Rg3 的 PEG-b-PPS 纳米颗粒可改善心脏功能并减少梗死面积。G-Rg3 靶向叉头框转录因子 3a（forkhead box 3a，FOXO3a），通过抑制氧化应激、炎症和纤维化过程来调节下游信号通路[87]。G-Rg3 可改善 BLM 诱导的小鼠肺纤维化，下调 TGF-β1、Smad2、Smad3、MMP-2、MMP-9 和金属蛋白酶组织抑制剂-1 的表达，上调 Smad7 的蛋白表达。这些结果提示 G-Rg3 对 BLM 诱导的肺纤维化的保护作用与 TGF-β1/Smad 信号通路和 MMP 系统的调节有关[88]。在 BLM 诱导小鼠肺纤维化模型中，G-Rg3 给药抑制 BLM 诱导的 HIF-1α 的核定位增加，从而抑制 TGF-β1/Smad 信号通路，进而抑制 EMT 介导的成纤维细胞过度增殖，减缓肺纤维化的进程[89]。DN 大鼠模型中，20(S)-G-Rg3 处理后肾组织病理学变化得到减轻，肾小管上皮细胞凋亡显著减少，尿蛋白降低。20(S)-G-Rg3 治疗组的空腹血糖、肌酐、总胆固醇和甘油三酯水平均低于糖尿病组。机制上，20(S)-G-Rg3 显著下调肾脏中 TGF-β1、NF-κB 65 和 TNF-α 的表达，预防炎症引起的肾损伤[90]。

在 TGF-β1 处理的人腹膜间皮细胞（human peritoneal mesothelial cell，HMrSV5）中，G-Rg3 给药抑制了 EMT、波形蛋白和 α-SMA 的表达及细胞迁移。G-Rg3 降低了 TGF-β1 诱导的 Akt 和 Smurf2 的激活。G-Rg3 通过抑制 Akt 的激活来抑制 HMrSV5 中由 TGF-β1 诱导的 EMT[91]。

10. 人参皂苷 Rk3

在 DMN 和 CCl$_4$ 诱导的肝细胞癌小鼠模型中，G-Rk3（图 3-16）的应用抑制了肝损伤、肝纤维化和肝硬化。同时，G-Rk3 通过降低炎症细胞因子的表达、诱导细胞凋亡和阻断细胞周期来降低炎症反应。G-Rk3 还能有效地改善肠道菌群失调。此外，相关性分析表明，被 G-Rk3 抑制的 LPS-TLR4 信号通路在预防肝细胞癌中起着关键作用[92]。

11. 20S-原人参三醇

原人参三醇的代谢物 20S-原人参三醇（图 3-17）可以抑制 ECM 沉积并

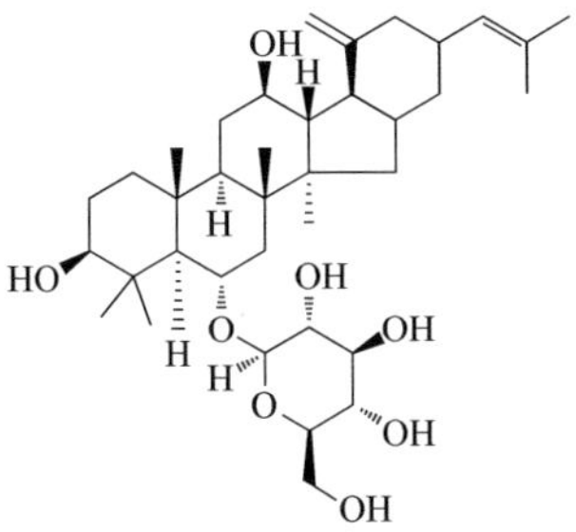

图 3-16 人参皂苷 Rk3 分子结构式

降低促炎细胞因子的水平，例如 casase 1、IL-1β、IL-1R1 和 IL-6，还显著增加法尼醇 X 受体（farnesoid X receptor，FXR）的表达，抑制嘌呤能配体门控离子通道 7 受体（P2X7 purinergic receptor，P2X7r）信号通路。在 TAA 诱导的 LF 小鼠中，20S-原人参三醇可以减弱组织病理学变化并调节 FXR 和 P2X7r 的表达水平。20S-原人参三醇活化 FXR，诱导 P2X7r 分泌减少、抑制 HSCs 活化并减轻炎症，最终减轻 LF[93]。

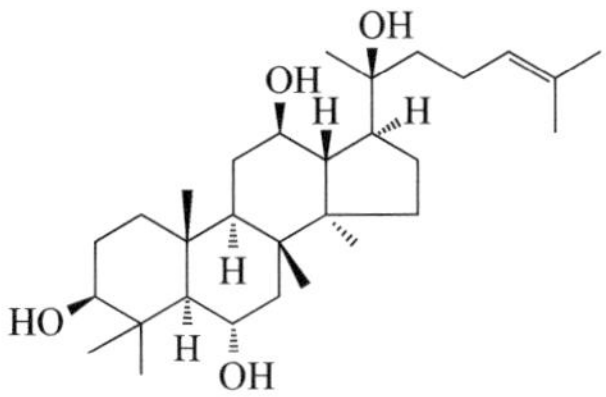

图 3-17 20S-原人参三醇分子结构式

12. 人参皂苷提取物

人参皂苷提取物在 FFD 诱导的 NAFLD 小鼠模型中发挥抗脂肪、抗纤维化和抗炎作用。研究发现，人参皂苷提取物富含 G-Rh1 和 G-Rg2，它们通过抑制 NLRP3 炎性体、促进线粒体自噬和减轻 mtROS 产生，从而对 NAFLD 发挥保护作用[94]。

第十一节
丹参酮ⅡA

丹参酮ⅡA（tanshinone ⅡA，Tan ⅡA，图3-18）由唇形科植物丹参的干燥根提取。

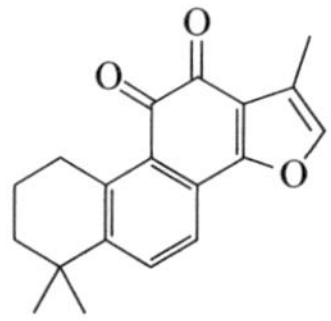

图3-18 丹参酮ⅡA分子结构式

1. 肝

在CCl_4处理诱导的大鼠肝损伤模型中，Tan ⅡA处理显著改善了肝脏的组织学损伤，降低白蛋白-胆红素（albumin-bilirubin，ALBI）评分和终末期肝病（end-stage liver disease，ESLD）模型评分。Tan ⅡA处理后改变了与新增殖的细胞ALB、OV-6、EPCAM和CK-18共定位的蛋白质的表达。Tan ⅡA可减轻CCl_4联合酒精所致大鼠肝损伤，并通过促进内源性肝干细胞的增殖和分化，在肝硬化中发挥治疗作用[95]。

2. 肺

Tan ⅡA可能通过抑制TGF-β1/Smad信号、抑制NOX4表达和激活Nrf2/ARE通路来保护肺免受SiO_2诱导的损伤和纤维化[96]。Tan ⅡA可以在体内和体外改善硅肺纤维化，下调COL-1、COL-3和α-SMA的表达。Tan ⅡA处理有效抑制TGF-β1诱导的Smads磷酸化，尤其是Smad3在细胞核中的持续磷酸化，并上调Smad7在硅肺成纤维细胞中的表达，减少

ECM 沉积。Tan ⅡA 通过抑制 TGF-β1-Smad 信号通路的激活来部分改善硅肺纤维化[97]。在硅肺大鼠模型中，用 Tan ⅡA 进行体内治疗可显著缓解 SiO_2 诱导的肺纤维化。Tan ⅡA 显著抑制 SiO_2 诱导的 EMT 和 TGF-β1/Smad 信号通路，减轻氧化应激并激活核因子 A549 和人支气管上皮（human bronchial epithelial，HBE）细胞中的 Nrf2 信号通路。总之，Tan ⅡA 通过调控 Nrf2 介导的 EMT 抑制和 TGF-β1/Smad 信号转导减轻 SiO_2 诱导的肺纤维化[98]。

3. 肾

在链脲佐菌素（STZ）诱导的大鼠糖尿病肾病模型中，Tan ⅡA 减弱了血清尿酸（serum uric acid，UA）和血尿素氮（blood urea nitrogen，BUN）水平的升高，并恢复了血清中 SOD 活性。Tan ⅡA 改善了糖尿病大鼠肾小球基底膜和肾小管增厚及胶原蛋白的沉积。Tan ⅡA 处理降低了糖尿病大鼠中 TGF-β1、血小板反应素-1（thrombospondin-1，THBS-1）、Grp78 和增强子结合蛋白同源蛋白的表达水平。Tan ⅡA 减轻了糖尿病大鼠肾组织中 p-PERK、p-elf2α 和激活转录因子 4（activated transcription factor 4，ATF4）蛋白水平的升高。Tan ⅡA 介导的 STZ 诱导的 DN 的治疗作用可能与通过抑制 PERK 信号通路而减轻的 ER 应激有关[99]。

4. 系统硬化

在 BLM 诱导的体内和体外的皮肤纤维化模型中，Tan ⅡA 通过改善皮肤厚度和胶原蛋白沉积具有很强的抗纤维化作用。此外，Tan ⅡA 抑制了 EMT，抑制 Akt/mTOR/p70S6K 通路的激活。总之，Tan ⅡA 减轻了系统硬化相关的皮肤纤维化和 EMT，并且 Akt/mTOR/p70S6K 信号通路参与了这一调节。因此，Tan ⅡA 具有作为治疗系统硬化血管损伤疾病药物的潜力[100]。

5. 肌腱

在体外成纤维细胞和大鼠跟腱模型中，miR-29b 抑制剂和 Tan ⅡA 联合给药可以激活 TGF-β/Smad3 通路以触发内源性通路并诱导成纤维细胞的高度增殖。在 miR-29b 处理 6h 后添加 Tan ⅡA，可使大鼠模型产生的细胞毒性更小，且肌腱粘连减少、强度增强的结果更好。Tan ⅡA 和 miR-29b 抑制剂联合治疗能够通过防止肌腱粘连和增强肌腱强度来改善肌腱愈合[101]。

第十二节
红景天苷

红景天苷（salidroside，图 3-19）是一种从红景天中分离出来的活性多酚。

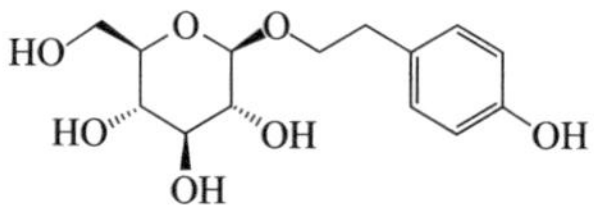

图 3-19 红景天苷分子结构式

1. 心

在冠状动脉结扎 4 周建立的兔心衰模型中，红景天和 SAL 治疗使白细胞（white blood cell，WBC）和 $CD4^+$ T 细胞水平下降，IL-17 及其下游靶基因 IL-6、TNF-α、IL-1β、IL-8 和 CC 趋化因子配体 2（CC chemokine ligand 2，CCL2）的表达降低，NLRP3 炎性体水平降低，纤维化和胶原蛋白生成显著减少，p38 MAPK 和 ERK1/2 磷酸化减弱，诱导性 VA 降低，Kir2.1、Nav1.5、NCX、PLB、SERCA2a 和 RyR 水平上调。红景天抑制 MAPK 的激活介导 IL-17 及其下游靶基因的表达，从而降低了纤维化和细胞凋亡的水平并抑制了室性心律失常[102]。在糖尿病心肌病小鼠模型中，SAL 治疗在 0.025mg/kg 和 0.05mg/kg 浓度下可防止心肌细胞凋亡和心室重构。血红素氧合酶-1（heme oxygenase-1，HO-1）参与 SAL 对糖尿病的有益抗氧化和心脏保护作用。SAL 的这种心脏保护作用依赖于 Akt/Nrf2/HO-1 信号激活，进而减轻 DCM[103]。在冠状动脉结扎建立的小鼠心肌梗死模型中，

SAL 显著降低死亡率、改善心脏功能、减少纤维化和梗死面积。此外，口服 SAL 可减轻心肌炎症和细胞凋亡，促进血管生成。SAL 下调 TNF-α、TGF-β1、IL-1β 和 Bax 的表达水平，上调 Bcl-2、VEGF、Akt 和 eNOS 的表达。这些结果表明，SAL 可以减轻 MI 小鼠心肌重塑的病理过程，可能是治疗临床缺血性心血管疾病的一种潜在有效的治疗方法[104]。

2. 肝

SAL 可减轻棕榈酸/油酸刺激的原代肝细胞中的脂质积累和炎症反应。此外，SAL 通过调节葡萄糖代谢失调、胰岛素抵抗、脂质积累、炎症和纤维化，有效预防高脂肪/高胆固醇饮食诱导的 NASH 进展。从机制上讲，SAL 在体外和体内促进了 AMPK 信号通路的激活，抑制肝细胞和肝脏脂质积累和炎症，进而减轻 NASH[105]。

3. 肾

在 UUO 或叶酸（folic acid，FA）诱导的小鼠体内肾间质纤维化模型和 TGF-β1 刺激的 HK-2 细胞体外模型中，SAL 治疗可以改善肾小管损伤和 ECM 的沉积。SAL 抑制 EMT，这可以通过 α-SMA、波形蛋白、TGF-β1、Snail、Slug 的表达降低和 E-cadherin 的大量恢复表达来证明。此外，SAL 还降低血清生化标志物 Scr、BUN、UA 的水平，并减少炎性细胞因子（IL-1β、IL-6 和 TNF-α）的释放。进一步的研究表明，SAL 对 RIF 的影响与体内和体外 TLR4、p-IκBα、p-NF-κB 和 MAPK 的低表达有关。总的来说，SAL 通过抑制 TLR4/NF-κB 和 MAPK 信号通路改善 RIF[106]。在 STZ 给药肥胖小鼠的糖尿病肾病动物模型中，SAL 可以减少蛋白尿、肾纤维化和足突消失，同时抑制 nephrin 和 podocin 的表达降低。SAL 可能通过 Sirt1/过氧化物酶体增殖活化受体 γ 共激活因子-1α（peroxisome proliferator-activated receptor gamma coactivator-1α，PGC-1α）轴调节线粒体生物发生，因此它可作为治疗 DN 的潜在候选药物[107]。在 ADM 诱导的肾病小鼠模型中，SAL 治疗可改善蛋白尿，抑制 nephrin 和 podocin 的表达，并减少由 ADM 引起的肾纤维化和肾小球硬化。从机制上讲，SAL 的应用在很大程度上消除了 β-catenin 的核转位，从而抑制其活性。总之，SAL 抑制 β-catenin 信号通路可改善 ADM 肾病中的蛋白尿、肾纤维化和足细胞损伤[108]。

4. 血管

在同型半胱氨酸（homocysteine，Hcy）预处理的 HUVECs 中，SAL 治疗上调了血管内皮钙黏蛋白（vascular endothelial-cadherin，VE-cadherin）的表达水平，下调了 α-SMA 的表达水平，减少了细胞迁移并激活了 eNOS/一氧化氮（nitric oxide，NO）信号轴，以及下调 Krüpple 样因子 4（krüpple-like factor 4，KLF4）表达和易位至细胞核。研究结果表明，SAL 可能通过调节 KLF4/eNOS 信号通路来抑制 Hcy 诱导的 EMT[109]。

5. 关节

在前交叉韧带横断（anterior cruciate ligament transection，ACLT）诱导的骨关节炎（osteoarthritis，OA）大鼠模型中，SAL 可显著促进软骨细胞增殖，恢复软骨组织学改变。SAL 上调 COL-2 和 Aggrecan 的水平，下调 MMP-13 的水平，减少 CD4$^+$ IL-17$^+$ 细胞的数量，降低 IL-17、IKBα 和 p65 的水平，同时提高 CD4$^+$ IL-10$^+$ 细胞的数量和 IL-10 的水平。IL-17 的减少进一步抑制了 IKBα 与 p65 的解离，从而减少了 TNF-α 和血管细胞黏附分子-1（vascular cell adhesion molecule-1，VCAM-1）的释放。总之，SAL 通过促进 ACLT 诱导的 OA 大鼠中的软骨细胞增殖、抑制胶原沉积和纤维化以及通过 NF-κB 通路调节炎症和免疫反应来减轻软骨退化[110]。

6. 海绵体

在双侧海绵体神经损伤（bilateral cavernous nerve injury，BCNI）的雄性大鼠模型中，SAL 治疗不仅改善了 BCNI 大鼠的勃起功能，而且通过促进保护性自噬，抑制了细胞凋亡和纤维化，改善了神经内容物、内皮细胞和海绵体平滑肌细胞的损失。SAL 对海绵神经损伤大鼠模型勃起功能和海绵体保护性自噬有增强作用[111]。

第十三节
姜黄素

姜黄素（curcumin，图 3-20）是从姜黄根中分离出来的一种低分子量的多酚类化合物，具有多种生物学活性。其在食品工业中被广泛用作天然色素。

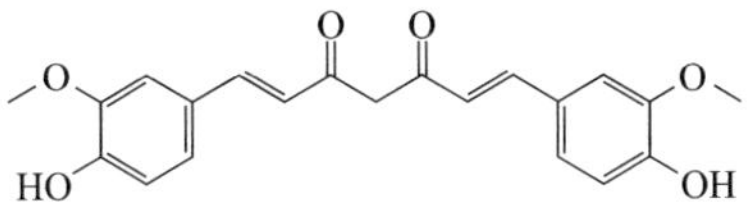

图 3-20 姜黄素分子结构式

1. 心

在 T2DMdb/db 小鼠中建立了糖尿病心肌病小鼠模型，姜黄素可以抑制 ROS 的产生，减少心肌细胞凋亡和心肌脂质积累。姜黄素上调了 Bcl-2 的表达，并下调了 Bax 和 Caspase-3 蛋白的表达。因此，姜黄素通过抑制 ROS 的产生对 DCM 具有一定的治疗作用[112]。

在 1 型糖尿病（type 1 diabetes mellitus，T1DM）相关的 DCM 中，吡格列酮和/或姜黄素治疗显著降低了血清心脏损伤标志物和血脂谱标志物水平，减轻了氧化应激和纤维化，显著抑制心肌脂质过氧化和 TGF-β1 水平，同时显著升高总抗氧化能力并改善心脏组织的组织病理学结构。吡格列酮/姜黄素治疗通过抑制心脏钙调蛋白依赖蛋白激酶Ⅱ（calmodulin-dependent protein kinaseⅡ，CaMKⅡ）/NF-κB 信号转导并伴随 PPARγ 表达增强来实现其对心脏的保护作用[113]。

在 Ach-CaCl$_2$ 诱导房颤大鼠模型中，姜黄素显著减少房颤持续时间并降低左心房纤维化程度，显著降低 IL-17A、IL-1β、IL-6 和 TGF-β1 的分泌。生物信息学分析表明，IL-17 信号通路参与了姜黄素的治疗机制。此外，编码 COL-1a1、Fasn、Pck1、Bmp10、IL-33 和 Figf 的基因是与姜黄素对心房颤动的治疗机制相关的关键基因[114]。

此外，姜黄素显著减弱了 DS 大鼠心脏中 GATA 结合蛋白 4（GATA-binding protein 4，GATA4）的乙酰化水平。姜黄素是一种 p300-HAT 活性抑制剂，可抑制 Dahl 大鼠中高血压引起的 LV 肥大的发展，并保留射血分数[115]。

2. 肝

姜黄素增加了活化的 HSCs 中脂滴（lipid droplets，LD）的形成，刺激了甾醇调节元件结合蛋白和脂肪酸合酶的表达，并降低了脂肪甘油三酯脂肪酶的表达。姜黄素通过增加体外活化的 HSCs 中 Perilipin5 基因的表达来恢复细胞内脂滴的形成。因此，姜黄素是潜在地治疗 LF 的安全有效的候选药物[116]。

在 CCl_4 诱导的 LF 中，姜黄素可减少肝损伤、氧化应激、纤维化，并恢复 MMP-9 和 MMP-2 的正常活性。此外，姜黄素可恢复 CCl_4 诱导的 NF-κB、IL-1、IL-10、TGF-β、CTGF、COL-1、MMP-13 和 Smad7 蛋白水平变化。另外，姜黄素降低 JNK 和 Smad3 磷酸化，并降低了 α-SMA 和 Smad3 蛋白以及 mRNA 水平。姜黄素通过调控谷胱甘肽、NF-κB、JNK-Smad3 和 TGF-β-Smad3 通路，使活化的 HSCs 减少，从而发挥抗纤维化作用[117]。

在 BDL 诱导的纤维化大鼠中，姜黄素、槲皮素和阿托伐他汀治疗导致 BDL 组中 miR-21 和 TGF-β1 的下调以及 miR-122 的上调。这些药物在改变基因表达方面没有显著差异，并且都具有相同的效果。此外，在 miR-21 和 TGF-β1 之间观察到直接显著相关性，并且在 miR-122 和 TGF-β1 表达之间观察到负显著相关性。总之，靶向这些分子途径可能部分阻止 LF 的进展[118]。

在 TAA 诱导的 LF 中，石榴提取物和/或姜黄素减轻 LF，如肝功能指标（AST、ALT、ALP 和 ALB）、氧化应激生物标志物（MDA、SOD 和 GSH）和炎症生物标志物［NF-κB、TNF-α、IL-1β、iNOS、TGF-β 和髓过氧化物酶（myeloperoxidase，MPO）］的显著改善。此外，PE 和/或姜黄素治疗显著上调 Nrf2/HO-1 基因表达，同时显著下调 NF-κB、TGF-β 和磷酸化 Smad3 蛋白表达，以及下调 α-SMA 和 COL-1 基因表达。总之，PE 和/或姜黄素的保肝活性可能与其调节 Nrf2/HO-1、NF-κB 和 TGF-β/Smad3 信号通路的能力有关。值得注意的是，与单一疗法相比，PE 和姜黄素的组合对 TAA 诱导的 LF 具有更好的保肝作用[119]。

随机、安慰剂对照的临床试验中，52 名 NAFLD 患者被随机分配接受生活方式建议加 1500mg 姜黄素或安慰剂治疗 12 周。姜黄素治疗组 LF、血清胆固醇、葡萄糖和 ALT 显著降低。两组人体测量指标、血脂、胰岛素抵抗和肝脂肪变性均显著降低（$p<0.05$），且两组间无显著差异。结果表明，每日摄入 1500mg 姜黄素并进行减重治疗在改善 NAFLD 患者心血管危险因素方面并不优于单独减重[120]。

在 CCl_4 注射诱导的肝硬化大鼠模型中，姜黄素组的 IL-10 浓度显著增加，TNF-α 和 TGF-1β 水平显著降低。肝组织的组织病理学检查表明，姜黄素治疗组几乎保留了肝组织的正常结构。总之，姜黄素通过调节免疫系统机制对抗外源性化学毒性来抑制 LF[121]。

在瘦素缺陷型肥胖小鼠中，姜黄素降低了瘦素诱导的甲硫氨酸腺苷转移酶 2A（methionine adenosyltransferase Ⅱ alpha，MAT2A）的表达。姜黄素处理抑制 JNK 信号，进而抑制 MAT2A 水平升高。姜黄素通过影响 $-2847bp$ 和 $-2752bp$ 之间以及 $-2752bp$ 和 $+49bp$ 之间的 MAT2A 启动子片段来抑制瘦素诱导的 MAT2A 启动子活性。姜黄素通过抑制 JNK 信号转导影响瘦素诱导的 HSCs 中 MAT2A 的表达，进而发挥姜黄素对肥胖高瘦素血症患者 LF 的抑制作用[122]。

在随机双盲安慰剂对照试验中，70 名 20～70 岁的肝硬化患者被随机分为两组，分别接受 1000mg/天姜黄素（$n=35$）或安慰剂（$n=35$），为期 3 个月。在这项初步研究中，观察到补充姜黄素在降低肝硬化患者的疾病活动评分和肝硬化严重程度方面的有益作用[123]。在另一项研究中，根据在 CLDQ 领域（例如疲劳、情绪功能、担忧、腹部症状和全身症状）、LDSI 2.0 领域（例如瘙痒、关节疼痛、右上腹部疼痛、白天睡觉、食欲下降、抑郁、害怕并发症、黄疸、财务障碍、使用时间的变化、性兴趣下降和性活动减少）和 SF-36 领域（例如身体机能，身体疼痛、活力、社会功能和心理健康）的评估得知，姜黄素改善了肝硬化患者的生活质量[124]。

在 TAA 诱导的肝硬化大鼠中，姜黄素治疗显示纤维化减轻，肝脏生物标志物水平显著降低，抗氧化酶 SOD 和谷胱甘肽水平升高、MDA 水平降低、CAT 活性和电解质稳态恢复。这些发现证实了姜黄素在肝硬化中的保护作用[125]。

在 BDL 诱导的慢性肝病模型中，姜黄素减弱肝组织学、转氨酶和纤维化标志物 mRNA 表达的变化。此外，姜黄素显著增加了对氧磷酶 1（paraoxonase1，PON1）的 mRNA 和蛋白质表达，以及促进 PON1 表达的基因的 mRNA，如特化蛋白 1（specificity protein 1，Sp1）、蛋白激酶 Cα（protein kinase Cα，PK Cα）、胆固醇调节元件结合蛋白 2（sterol-regulatory element binding protein 2，SREBP-2）、芳香烃受体（aryl hydrocarbon receptor，AhR）、JNK，并通过上调载脂蛋白 A1（apolipoprotein A1，APO A1）增加 PON1 活性。姜黄素治疗可通过增加对 PON1 表达和活性的影响来抑制肝硬化进展[126]。

HSCs 激活是 LF 的关键步骤，需要重塑 DNA 甲基化，这与由催化亚基甲硫氨酸腺苷转移酶 MAT2A 和调节亚基 MAT2B 组成的蛋氨酸腺苷转移酶Ⅱ相关。姜黄素通过中断 p38 MAPK 信号通路降低 HSCs 中 MAT2B 的表达，进而抑制小鼠肝脏中 HSCs 活化和胶原蛋白的表达。姜黄素通过阻断 HSCs 中 p38 MAPK 的磷酸化来降低 MAT 2B 的表达，进而发挥抗纤维化作用[127]。

在双酚 A（bisphenol A，BPA）诱导啮齿动物的 LF 模型中，姜黄素、N-乙酰半胱氨酸和蜂胶提取物三重组的干预措施显著减轻了肝损伤和纤维化，恢复了促氧化/抗氧化平衡，将细胞因子平衡转向抗炎侧，降低 IL-1β/IL-10 的比例。联合治疗通过增加肝细胞中 Bcl-2 表达和降低肝 Caspase-3 含量来发挥抗凋亡作用，通过下调 MMP-9 和上调 MMP-2 基因表达改善 ECM 增加。联合治疗可防治 BPA 诱导的 LF，因为它们具有抗氧化、抗炎和抗凋亡特性，可减少 ECM 的更新[128]。

在 CCl_4 诱导的仓鼠肝硬化模型中，与同时服用姜黄素相比，单独服用多沙唑嗪或卡维地洛显示出较弱的逆转肝硬化的能力。然而，效果最好的是多沙唑嗪、卡维地洛和姜黄素的联合给药，可以逆转 LF 并减少 COL-1 的含量，显示肝功能正常（葡萄糖、ALB、AST、ALT、TB 和 TP）。此外，三种药物联合给药增加了肝脏炎症细胞中 Nrf2/NF-κB mRNA 的比例及其蛋白质表达，这可能是另一种保肝机制。因此，含有姜黄素的 α/β 肾上腺素能阻滞剂可完全逆转肝损伤，这可能是由于肾上腺素能拮抗 HSCs 的结果，并且可能是与 Nrf2/NF-κB mRNA 比例的增加有关[129]。

在果糖（20%）饮水诱导的青少年 NASH 中，姜黄素可以减轻炎症，下调 AMPKα 的肝脏基因表达比例，上调 TNF-α 的表达比例。新生儿姜黄素给药是预防 NASH 的潜在天然药理学候选物[130]。

在 HFD 诱导的 NASH 中，姜黄素作为预防性或治疗性给药可显著保留和/或恢复肝脏结构，这通过转氨酶的正常化、细胞变化的保存和/或可逆性以及纤维化阶段的降低得到证明。姜黄素通过激活 Nrf2 发挥对 NASH 的预防和治疗作用[131]。

在一项双盲、随机、安慰剂对照试验中，80 名 NAFLD 患者被随机分配接受每天 250mg 的姜黄素或安慰剂治疗 2 个月，与安慰剂相比，姜黄素组的肝脂肪变性程度和血清 AST 水平显著降低。两组的高密度脂蛋白（high density lipoprotein，HDL）水平和人体测量学指标也显著降低，两组之间没有显著差异。与安慰剂相比，每天补充低剂量磷脂姜黄素的 NAFLD 患者的肝脂肪变性和相关酶活性显著降低[132]。

在 CCl_4 诱导的肝纤维化模型中，姜黄素通过激活自噬来抑制 TGF-β/Smad 信号转导，从而抑制 EMT。Smad2 和 Smad3 降解性多泛素化的机制可能是通过抑制四肽重复结构域 3，并诱导 Smad 泛素化调节因子 2（smad ubiquitination regulatory factor 2，Smurf2）发生泛素化和蛋白酶体降解。姜黄素通过激活 PPARα 抑制肝细胞中的 ROS 水平和氧化应激，并调节 AMPK 和 PI3K/Akt/mTOR 的上游信号通路，导致肝细胞中的自噬流增加。姜黄素通过激活自噬有效减少了肝细胞中 EMT 的发生并抑制了 ECM 的产生，这为治疗 LF 提供了一种潜在的新策略[133]。

3. 肺

在 BLM 或 TGF-β1 处理的细胞中，蛋白质组学分析显示姜黄素逆转了特定蛋白质的表达，如 DNA 拓扑异构酶 2α（topoisomerase 2α，TOP 2A）、驱动蛋白家族成员 11（kinesin family number 11，KIF11）、着丝粒蛋白 F (centromere protein F, CENPF) 等。姜黄素可恢复 TGF-β1 诱导的过氧化物酶体的表达，如 PEX-13、PEX-14、PEX-19 和 ACOX1。姜黄素在急性肺损伤体外模型中的独特分子靶点与抗氧化和抗纤维化活性相关[134]。

在 BLM 诱导的 C57BL/6 小鼠肺纤维化模型中，姜黄素治疗抑制小鼠体重减轻和肺形态改变，并与 p53、尿型纤溶酶原激活剂（uroplasminogen activato, uPA）和纤溶酶原激活抑制剂（plasminogen activator inhibitor, PAI）蛋白有强相互作用。姜黄素靶向 IL-17A 介导的 p53-纤溶系统，从而发挥治疗肺纤维化的作用[135]。

在 A549 细胞系和 BLM 诱导的小鼠肺纤维化模型中，姜黄素下调细胞增殖相关抗原（cell proliferation-associated nuclear antigen Ki67, Ki67）和表皮生长因子受体（epidermal growth factor receptor, EGFR）的表达水平。姜黄素在体外和体内抑制急性肺损伤和肺纤维化中的 EGFR 和 Ki67，进而发挥抗纤维化作用[136]。

在 BLM 诱导的肺纤维化模型中，姜黄素显著抑制了支气管肺泡灌洗液（bronchoalveolar lavage fluid, BALF）和肺 FN 水平的增加。用姜黄素治疗 BLM 大鼠可显著抑制 AMs 释放 FN，抑制肺中复合碳水化合物和糖苷酶的增加。这些发现表明，BLM 诱导的肺纤维化与糖蛋白的积累有关，而姜黄素具有抑制纤维化肺中糖蛋白沉积的能力[137]。

在 TGF-β1 诱导人肺成纤维细胞（human lung fibroblasts, HLF）过度增殖模型中，姜黄素或姜黄醇的使用显著降低了 HYP 和 α-SMA 的水平，并且减少了 COL-1 和 COL-3 的沉积。此外，与胶原合成相关的蛋白质包括Ⅰ型胶原的 N 端前肽（procollagen type Ⅰ N-terminal peptide, PⅠNP）、Ⅲ型胶原的 N 端前肽（procollagen type Ⅲ N-terminal peptide, PⅢNP）和脯氨酰羟化酶（prolyl hydroxylase, PHD）均以剂量依赖性降解。在 TGF-β1 诱导的 HLF 过度增殖模型中，自噬作为清除剂的作用被削弱，姜黄素或姜黄醇处理增加细胞质中自噬体，上调酵母 ATG 6 同系物（mammalian ortholog of yeast ATG 6, Beclin1）和自噬相关基因 7（autophagy-related gene 7, ATG7）。姜黄素或姜黄醇通过自噬减少 HLF 中胶原蛋白的沉积[138]。

在 BLM 诱导的急性肺损伤模型中，姜黄素可阻断 COX-2、NF-κB-p65、FN 的表达，并在体内表达 P-AMPKα。姜黄素还可以抑制 BLM/IL-17A 暴露小鼠中 NF-κB-p105 的表达。mRNA 表达显示姜黄素处理后血小板衍生生长因子 A（platelet-derived growth factor subunit A，PDGFA）、PDGFB、CTGF、胰岛素样生长因子 1（insulin-like growth factor 1，IGF1）、NF-κB1、NF-κB2、MMP-3、MMP-9 和 MMP-14 的表达降低。研究表明，姜黄素通过 AMPKα/COX-2 途径发挥作为治疗肺损伤的辅助抗炎药物的潜在作用，在肺纤维化的发生发展中发挥关键作用[139]。

在 BLM 诱导的大鼠肺纤维化模型中，姜黄素通过调节肺中的胶原更新、组装和沉积来防止纤维化沉积，并且姜黄素治疗通过抑制 TGF-β1 的释放来防止巨噬细胞的 BLM 活化，进而发挥抗纤维化作用[140]。

H_2O_2 诱导的小鼠肺间充质干细胞（lung mesenchymal stem cells，LMSCs）损伤模型中，姜黄素处理组的 ROS 水平降低，而线粒体膜电位水平呈浓度依赖性升高，裂解型 caspase-3 表达降低，Nrf2 和 HO-1 表达增加，Bcl-2/Bax 和 p-Akt/Akt 比值增加。姜黄素对 H_2O_2 介导的小鼠 LMSCs 损伤的保护作用可能是通过 Akt/Nrf2/HO-1 信号通路介导的[141]。

在 PQ 暴露前用姜黄素预处理 A549 细胞 3h 可维持 E-cadherin 表达并抑制 PQ 诱导的 α-SMA 表达。姜黄素预处理有效抑制 TGF-β 表达，降低了 MMP-9 表达。姜黄素可以通过调节 TGF-β 的表达来调节 PQ 诱导的 EMT[142]。

CCD-19Lu 细胞是一种 HLF，在 CCD-19Lu 细胞中，姜黄素抑制 α-SMA 和成熟 TGF-β1 的表达，并抑制细胞分化。姜黄素诱导组织蛋白酶 B（cathepsins B，CatB）和组织蛋白酶 L（CatL）的显著上调，胱抑素 C 被下调，这使得分泌的组织蛋白酶的肽酶活性恢复并恢复蛋白水解平衡。同时，不溶性 COL-1 和可溶性 COL-1 的水平均降低，达到与未分化成纤维细胞观察到的水平相似的水平。姜黄素触发 PPARγ 的表达。姜黄素通过 PPARγ 上调 CatB 和 CatL 抑制 TGF-β1 依赖性肺成纤维细胞分化[143]。

在 IPF 的成纤维细胞中，姜黄素降低 hsa-miR-6724-5p 的水平，致使 KLF10 表达增加，从而导致姜黄素处理的 IPF 成纤维细胞中的细胞周期停滞。总之，姜黄素在 IPF 治疗中发挥潜在作用[144]。

4. 肾

在 5/6 肾切除术诱导的大鼠慢性肾功能衰竭（chronic renal failure，

CRF）模型中，姜黄素抑制血液生化指标、蛋白尿和肾指标升高，并且能够抑制 mTOR/HIF-1α/VEGF 通路激活。该结果表明，姜黄素通过阻断 mTOR/HIF-1α/VEGF 信号通路的激活在 CRF 的发生和发展中起重要作用，从而发挥肾脏保护作用[145]。

在 UUO 大鼠的 RIF 模型中，姜黄素处理使肾纤维化减弱，NLRP3 炎性体活化受到抑制，线粒体功能障碍得到改善，LC3B/LC3A 比值和 Beclin-1 表达增加。此外，姜黄素抑制 PI3K/Akt/mTOR 通路。这些结果表明姜黄素是一种很有前途的 RIF 治疗剂，其抗纤维化作用可能是调节 UUO 大鼠的自噬和保护线粒体功能来抑制 NLRP3 炎性体活性[146]。

在顺铂诱导的肾纤维化模型中，三氧化二砷（arsenic trioxide，As_2O_3）[3.5mg/(kg·d)，口服] 和姜黄素 [200mg/(kg·d)，口服] 抑制了肾功能标志物改变（如肌酐清除率和尿素氮水平）并改善纤维化（如降低 TGF-β1 mRNA 水平、α-SMA 蛋白水平和 HYP 含量）。As_2O_3 和姜黄素阻断顺铂激活的刺猬信号通路（hedgehog signaling pathway，Hh），表现为 Shh、Smo 和 Ptch 的 mRNA 水平降低，肾胶质瘤相关癌基因-1 和 Gli2 蛋白水平受到抑制。结果表明，As_2O_3 和姜黄素具有治疗纤维化的作用，并表明阻断 Hh 信号转导可能是缓解肾纤维化的有效方法[147]。

在手术诱导 UUO 建立的 RIF 模型中，姜黄素显著降低血清 SCr 和 BUN 水平，并改善 UUO 诱导的大鼠肾小管损伤。姜黄素显著下调 TGF-β1、P-Smad2/3、Cleaved caspase-3、Cleaved caspase-8 和排斥性引导分子 B（repulsive guidance molecule B，RGMB，又称 Dragon）水平。总之，姜黄素下调 TGF-β1/Smad 信号通路并抑制大鼠和 HK-2 细胞中的 Dragon 和纤维化分子的产生[148]。

在腺嘌呤和氧嗪酸钾诱导的尿酸肾病（uric acid nephropathy，UAN）大鼠模型中，姜黄素治疗可降低血清尿酸、肌酐和 BUN 水平。治疗组的肾脏病理损伤和代谢性内毒素血症也有所减轻，并改善了紧密连接蛋白的表达。此外，姜黄素改变了 UAN 大鼠的肠道微生物群结构，防止 UAN 中机会性病原体的过度生长，包括大肠杆菌-志贺氏菌和拟杆菌，并增加产生短链脂肪酸（short-chain fatty acid，SCFA）的细菌的相对丰度，例如乳酸杆菌和瘤胃球菌科。这些结果表明姜黄素可以调节肠道微生物群，强化肠道屏障，减轻代谢性内毒素血症，从而保护 UAN 大鼠的肾功能[149]。

在腹腔注射乙醛酸盐（100mg/kg）建立的肾结石模型中，姜黄素可显著减少草酸钙（Ca-oxalate，CaOx）沉积和同时发生的小鼠肾脏组织损伤，降低 MDA 含量和增加 SOD、CAT、GPx、GR 和谷胱甘肽水平来缓解氧化应激反应。此外，姜黄素治疗显著抑制高草酸尿诱导的细胞凋亡和自噬。姜黄素还减弱了 IL-6、MCP-1、骨桥蛋白（osteopontin，OPN）、CD44、α-SMA、COL-1 和胶原原纤维沉积的高表达。姜黄素治疗抑制小鼠肾脏中 Nrf2 及其主要下游产物 HO-1、NQO1 和 UGT 的总表达和核积累降低。姜黄素可显著减轻小鼠肾脏 CaOx 晶体沉积及并发肾组织损伤。潜在机制涉及抗氧化、抗凋亡、抑制自噬、抗炎和抗纤维化活性，并通过 Nrf2 信号通路降低 OPN 和 CD44 表达发挥多效抗结石特性[150]。

5. 内皮

在经 TGF-β1 处理的 HUVECs 中，姜黄素显著减弱 TGF-β1 的作用，同时增加 VE-cadherin、二甲基精氨酸二甲胺水解酶 1（dimethylarginine dimethylamino hydrolase1，DDAH1）和 Nrf2 的水平，降低 MMP-9 和 ERK1/2 磷酸化。结果表明，姜黄素通过 Nrf2 通路刺激 DDAH1 表达来抑制 TGF-β1 诱导的 EMT，从而减轻内皮细胞纤维化[151]。

6. 口腔

在 40 名临床诊断为口腔黏膜下纤维化（oral submucous fibrosis，OSMF）的患者中，A 组接受姜黄素凝胶，B 组接受姜黄素黏附贴剂，姜黄素凝胶和黏附贴剂能有效改善 OSMF 患者张口和减轻烧灼感。因此，它们可以被认为是治疗 OSMF 的安全、无创方式[152]。

在另一项临床研究中，当同时以全身和局部形式给药时，姜黄素在治疗口腔黏膜下纤维化方面的效果优于单独全身给药或抗氧化剂。姜黄素有可能成为传统处方药的有效替代品[153]。

将 120 例确诊为 OSMF 的患者随机分为Ⅰ、Ⅱ、Ⅲ组。Ⅰ、Ⅱ和Ⅲ组中的每位患者分别给予专业制备的姜黄素黏附半固体凝胶、曲安奈德和透明质酸酶黏附半固体凝胶的组合以及三者联合给药。与其他两组相比，三种药物联合给药实现了最大的张口度。与其他两组相比，曲安奈德和透明质酸酶组报告的视觉模拟评分法（visual analogue scale，VAS）烧灼感降低。与组Ⅰ（1%姜黄素）和组Ⅱ相比，组Ⅲ（1%姜黄素、1%透明质酸酶和0.1%曲安奈德联合）药物治疗显示出更好的黏膜颜色变化。姜黄素对诊断为OSMF的患者具有治疗作用，并在其他两种活性药物（即曲安西龙和透明质酸酶）的组合帮助下，实现了最大利用和增强的药物递送[154]。

对90名OSMF患者进行了一项随机安慰剂对照的平行临床研究，与安慰剂相比，姜黄素和番茄红素治疗组的临床结果都有统计学意义的改善。然而，同时发现姜黄素和番茄红素的治疗效果在OSMF患者中几乎相同[155]。

在采用简单随机化技术对确诊的OSMF病例进行平行组试验设计研究中，将患者分为两组，一组（30例）给予姜黄素凝胶，另一组（30例）给予芦荟凝胶，姜黄素凝胶和芦荟凝胶对改善OSMF症状有效，而芦荟凝胶对改善烧灼感更有效，且无任何副作用。因此，这两种药物可以作为推荐治疗之外的辅助治疗[156]。

7. 硬膜

在椎板切除术的大鼠EF模型中，局部施用姜黄素（100mg/kg）浸泡的可吸收明胶海绵可减少EF的形成，姜黄素的这种作用被认为是通过降低巨噬细胞、中性粒细胞和成纤维细胞等炎症细胞的功能以及抗炎和抗氧化作用来介导的[157]。

在大鼠椎板切除模型中，全身施用姜黄素可有效减少椎板切除区的EF形成、炎症、肉芽组织形成、脊髓回缩和异物反应。姜黄素的抗炎活性有利于减少EF的形成[158]。

8. 腹膜

在雄性SD大鼠连续8周灌输20mL 4.25%葡萄糖基PD标准液建立的腹膜纤维化模型中，姜黄素治疗显著增加超滤量，减少葡萄糖的传质和腹膜厚度。姜黄素可显著降低PD大鼠腹膜液中升高的TGF-β1，减弱PD大鼠腹膜中TGF-β1、α-SMA和COL-1蛋白和mRNA的增加。姜黄素可降低PD大鼠腹膜中TGF-β激活激酶1（TGF-β-activated kinase 1，TAK1）、JNK和

p38 的水平以及抑制 p-TAK1、p-JNK 和 p-p38 蛋白质的表达。结果表明，姜黄素对 PD 相关的腹膜纤维化具有保护作用，TAK1、p38 和 JNK 通路在其中发挥重要作用[159]。

在以葡萄糖为基础的腹膜透析液诱导 HMrSV5 中，姜黄素增加了细胞活力并抑制了细胞迁移。姜黄素处理导致 E-cadherin（上皮标志物）表达增加，α-SMA（间充质标志物）表达降低，降低了两种 ECM 蛋白、COL-1 和 FN 的 mRNA 表达，同时还降低了细胞中 TGF-β1 mRNA 和上清液 TGF-β1 蛋白的含量。此外，姜黄素显著降低了 p-TAK1、p-JNK 和 p-p38 的蛋白质表达。姜黄素通过抑制 HMrSV5 细胞 EMT 而干预纤维化进程，提示 TAK1、p38 和 JNK 通路在介导姜黄素对间充质细胞 EMT 的作用中具有重要意义[160]。

9. 脾

在长期高脂饮食模型中，姜黄素给药显著降低了脾脏中的铁水平。使用姜黄素补充剂后，由较高水平铁积累产生的脾组织炎症反应、细胞凋亡和纤维化得到改善。因此，铁补充剂与姜黄素补充剂一起使用应少于 4 个月，以避免在健康器官中出现额外积聚铁现象[161]。

10. 神经

在大鼠脊髓损伤的实验模型中，一种高度水溶性的纳米制剂姜黄素治疗发挥保护脊髓白质组织的作用，神经胶质瘢痕面积减少并且新发芽的轴突的数量增加。纳米配方姜黄素（lipodisq™）可调节局部炎症反应，减少神经胶质瘢痕并保护大鼠脊髓损伤后的白质[162]。

11. 囊性纤维化

在囊性纤维化支气管上皮细胞系 CFBE41o 细胞中，姜黄素处理可强烈抑制 TLR2 基因和蛋白质表达，显著降低了金黄色葡萄球菌肽聚糖依赖性 IL-8 基因上调。在原代人囊性纤维化支气管上皮细胞中也观察到 TLR2 基因表达和功能的强烈降低，但在人非囊性纤维化原代细胞中没有观察到。另外，姜黄素处理降低了转录因子 SP1 的核表达，该因子对于增加囊性纤维化细胞系和原代细胞中的基础 TLR2 表达至关重要。最后，抗氧化剂 N-乙酰半胱氨酸和蛋白酶体抑制剂 MG-132 减少了姜黄素依赖性 SP1 的减少，这表明氧化和蛋白酶体降解途径的关键作用。总之，研究表明姜黄素可能通过氧化过程加速 SP1 降解来下调和减弱囊性纤维化支气管上皮细胞中 TLR2 基因的表达和功能[163]。

12. 胆

姜黄素治疗在体外和体内显著减轻胆管上皮细胞（biliary epithelial cells，BECs）的 EMT。从机制上讲，姜黄素显著减弱了 TGF-β1 诱导的 Smad 和 Hedgehog 信号转导，并上调了 BEC 中的 CD109 表达。总的来说，姜黄素通过上调 CD109 的表达来减轻 BECs 的 EMT 进而减轻纤维化和小叶间胆管损伤[164]。

13. 唾液腺

在放射性碘（radioactive iodine，RAI）诱导的唾液腺（salivary gland，SG）功能障碍小鼠模型中，姜黄素或氨磷汀治疗恢复了唾液流速降低和延迟时间延长的情况。治疗组富含黏蛋白的腺泡较多，导管周围纤维化较少，组织发生重塑的迹象，表现为唾液上皮细胞（AQP-5 阳性）、SG 导管细胞（CK18 阳性）、内皮细胞（CD31 阳性）和肌上皮细胞（α-SMA 阳性）增多。治疗组减轻了 RAI 诱导的细胞死亡，表现出抗细胞凋亡作用，SOD 活性和超氧化物歧化酶 2（superoxide dismutase 2，SOD2）蛋白表达水平升高。研究结果表明，姜黄素可改善 RAI 诱导的小鼠 SG 功能障碍[165]。

14. 气道

在卵清蛋白诱导的过敏性哮喘小鼠模型中，鼻内姜黄素给药可抑制 BALF 中 MMP-2 和 MMP-9 的活性。而 MMP-9、组蛋白去乙酰化酶 1（histone deacetylasel，HDAC 1）、H3acK9 和 NF-kB p65 的蛋白质表达在姜黄素和丁酸钠预处理组中受到调节。鼻内姜黄素和丁酸钠通过抑制过敏性哮喘中的 HDAC，从而调节气道炎症和纤维化[166]。

15. 尿道

体外培养人尿道瘢痕成纤维细胞（human urethral stricture fibroblasts，HUSFs），姜黄素在体外增强了 HUSFs 的放射敏感性。此外，姜黄素和放射治疗促进了 HUSFs 的凋亡并阻断了 G2/M 期的细胞，通过 Smad4 途径抑制 COL-1 和 COL-3 的合成，并且可能参与自噬。结果表明，姜黄素可能是 HUSFs 的放射增敏剂，通过自噬下调 Smad4 抑制 HUSFs 的增殖和纤维化[167]。

16. 眼眶

眼眶纤维化是甲状腺相关性眼病（thyroid-associated ophthalmopathy，TAO）中组织重塑的标志，是一种慢性进行性眼眶病，几乎没有有效的治疗方法。原代细胞培养的三名 TAO 患者的眼眶成纤维细胞中，姜黄素处理抑制了 TGF-β1 信号通路，并减弱了 TGF-β1 诱导的 CTGF 和 α-SMA 的表达。姜黄素可降低 TGF-β1 诱导的眼眶成纤维细胞中的 MF 分化和促血管生成活性。研究结果支持姜黄素在治疗 TAO 中的潜在应用的观点[168]。

17. 新剂型姜黄素

短链聚乙二醇修饰的姜黄素衍生物 Curc-mPEG454 不仅增加了姜黄素的血药浓度，而且保留了其抗炎活性。在 CCl_4 诱导的大鼠 LF 模型中，50mg/kg 和 100mg/kg Curc-mPEG454 治疗显著降低了 ALT 和 AST 的升高，肝硬化的发生率从 75% 下降到 37% 和 35%。RNA-seq 分析显示，Curc-mPEG454 显著上调醛氧化酶 1（aldehyde oxidase 1，AOX1），同时下调细胞色素 P450 家族 26 亚家族 A 成员 1（cytochrome P450 26A1，CYP26A1）和 CYP26B1，从而恢复肝脏视黄酸水平，增加谷氨酸-半胱氨酸连接酶催化亚基（glutamate cysteine ligase catalytic subunit，GCLC）和谷氨酸-半胱氨酸连接酶修饰亚基（glutamate cysteine ligase modifier subunit，GCLM）表达合成肝内谷胱甘肽，并通过下调前列腺素 E 合酶 2（prostaglandin E synthase 2，PTGES2）/前列环素 E2（prostaglandin E 2，PGE2）信号转导抑制肝脏炎症。Curc-mPEG454 有效抑制受损肝脏中瘢痕相关巨噬细胞亚群和产生瘢痕的 MF 的扩增，并通过调节配体-受体相互作用（包括血小板衍生生长因子-B）重塑纤维化生态位/血小板衍生生长因子受体-α 信号转导。作为多靶点前药，聚乙二醇化姜黄素值得进一步关注和研究[169]。

在 STZ 诱导的 T1DM 的 C57BL/6 小鼠心脏和肾脏损害模型中，姜黄素和壳聚糖包裹的姜黄素治疗降低了血糖和总胆固醇水平，并增强了胰岛素分泌。然而，血压、甘油三酯含量和极低密度脂蛋白胆固醇含量没有变化。组织化学分析表明，姜黄素和壳聚糖包裹的姜黄素都改善了心脏 LV 的细胞肥大和核增大，并减少了肾脏的纤维化，尤其是在壳聚糖包裹的姜黄素治疗后。研究表明，壳聚糖能有效增强姜黄素对 T1DM 小鼠模型心脏和肾脏损害的治疗作用[170]。抗死亡受体 5（death receptor 5，DR5）抗体-姜黄素偶联物（anti-DR5 antibody-curcumin conjugate，DCC）是一种新型的复合物，其合成过程是通过马来酰亚胺功能化姜黄素衍生物和硫醇化抗 DR5 抗体之间的偶联反应合成。在单独抗 DR5 抗体和 DCC 之间未观察到活化 HSCs 的摄取行为有显著差异。由于姜黄素的抗氧化和抗炎作用，与单独抗 DR5 抗体处理的 HSCs 相比，经 DCC 处理的 HSCs 产生的 ROS 和诱导型一氧化氮合酶（inducible nitric oxide synthase，iNOS）水平低得多。此外，如 α-SMA 的免疫细胞化学分析所示，DCC 对活化的 HSCs 的抗纤维化作用比单独抗 DR5 抗体更显著。DCC 在对 LF 小鼠全身给药后优先在肝脏中积累。因此，DCC 可作为治疗 LF 的潜在治疗剂[171]。

负载姜黄素和环巴胺的脂质体有助于溶解这两种药物，有助于细胞吸收，抑制 HSCs 活化。载有这两种药物的脂质体抑制细胞增殖、迁移和侵袭，并诱导更多的细胞凋亡和扰乱细胞周期，并可以强烈抑制 HSCs 活化和胶原蛋白分泌。因此，脂质体包封可以增加姜黄素和环巴胺的摄取以及它们在抗纤维化中的协同作用[172]。

通过将铂与姜黄素缀合形成姜黄素铂纳米粒子 C-PtNPs 提高了姜黄素治疗 LF 的能力[173]。

在长期感染克氏锥虫（Trypanosoma cruzi）的小鼠模型中，使用次优剂量的标准杀寄生虫药物苯硝唑（benzonidazole，BNZ）与载有姜黄素的纳米颗粒联合进行口服治疗能够减少心肌寄生虫负荷、心脏肥大、炎症和纤维化。BNZ 和姜黄素联合治疗在下调心肌促炎细胞因子/趋化因子（IL-1β、

TNF-α、IL-6、CCL5）的表达以及 MMP-2、MMP-9 的水平及活性方面非常有效，同时也抑制了参与白细胞募集和心脏重塑的诱导酶（环氧合酶、一氧化氮合酶）的表达。因此，口服姜黄素的纳米制剂显示出作为传统治疗慢性南美锥虫心脏病药物 BNZ 的补充药物的潜力[174]。

第十四节
白藜芦醇

白藜芦醇（resveratrol，Res，图 3-21）是一种天然的抗氧化剂，是一种蒽醌萜类化合物，主要来源于蓼科植物虎杖（*Polygonum cuspidatum* Sieb. et Zucc.）的根茎提取物。它具有多种生物学作用，如抗肿瘤、保护心血管、抗氧化、抗自由基、抗菌、抗衰老以及具雌激素样作用和免疫调节作用等。

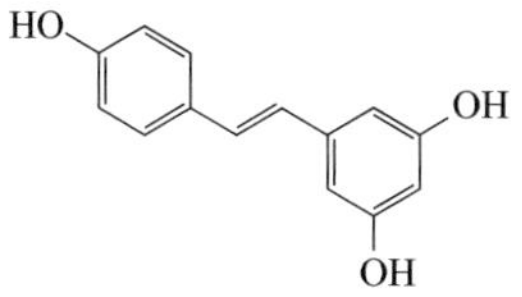

图 3-21　白藜芦醇分子结构式

1. 心脏

HFD 和 STZ 腹腔注射诱导糖尿病心肌病模型中，Res 改善了胰岛素抵抗，降低了甘油三酯、胆固醇和低密度脂蛋白胆固醇水平，逆转 DCM 大鼠受损的舒张和收缩心脏功能，改善心肌结构紊乱和纤维化，保留线粒体膜电位水平，抑制心肌细胞凋亡。Res 治疗可提高 DCM 大鼠心肌线粒体呼吸酶活性，同时减少 ROS 的产生。进一步研究表明，Res 处理进一步增加了 DCM 大鼠心肌和 HG 培养的 H9C2 心肌细胞中解偶联蛋白 2（uncoupling protein 2，UCP2）的表达。总之，Res 治疗改善了心脏功能并抑制了心肌细胞凋亡，其作用与 UCP2 介导的糖尿病大鼠的线粒体功能改善有关[175]。

通过横向主动脉缩窄（transverse aortic constriction，TAC）诱导的压力超负荷建立的小鼠心脏肥大重塑模型中，Res 的施用显著抑制了心脏肥大、纤维化和细胞凋亡，并改善了小鼠的体内心脏功能。Res 还逆转了心肌肥大并恢复了由慢性压力超负荷引起的收缩功能障碍。此外，Res 治疗阻断了 TAC 诱导的免疫蛋白酶体活性和催化亚基表达（β1i、β2i 和 β5i）的增加，从而抑制 PTEN 降解，进而导致 Akt/mTOR 失活和 AMPK 信号激活。总之，Res 是一种新型的免疫蛋白酶体活性抑制剂，可能是治疗肥厚性疾病的有前途的治疗剂[176]。

在肾大部切除诱导肾功能不全的慢性肾病模型中，Res 抑制 Scr 和 BUN 升高、肾小球硬化、肾小管间质纤维化，抑制肌细胞横截面积增加、间质和血管周围纤维化以及 LV 扩张。从机制层面讲，Res 治疗导致肾脏中 Mn SOD 的含量上调。Sirt1 是 Res 在慢性肾脏病啮齿动物模型中对肾脏和心脏损伤的保护作用的关键分子，而 FOXO1 和 Mn SOD 可能是 Sirt1 的下游靶标。因此，Sirt1 可能是治疗 CKD 引起的 LV 重构的潜在治疗靶点[177]。

Res 治疗降低了 CF 和 MF 的增殖。Res 治疗还以剂量依赖性方式增加了死亡 CF 和 MF 的比例。然而，用 Res 治疗不会诱导成年心肌细胞的细胞死亡。在 CF 和 MF 中，用 Res 处理的具有浓缩核的细胞百分比增加，但在心肌细胞中没有。结果表明，Res 治疗可以抑制大鼠 CF 和 MF 的细胞增殖并诱导细胞死亡，同时不影响心肌细胞的存活，且 CF 死亡与雌激素受体 α（estrogen receptor α，ERα）信号转导无关[178]。

HG 培养的 MSCs 中，Res 通过抑制存活标志物、激活 AMPK/Sirt1 轴和凋亡标志物的表达来降低 MSCs 的能力。自体脂肪干细胞移植糖尿病大鼠模型中，Res 通过 Sirt 1 和 IGFIR 表达增加 HG 应激下的 ADSCs 活力。Res 减轻纤维化、肥大和细胞凋亡，并激活 AMPK/Sirt1 轴。总之，MSCs 预处理的 Res 可能在治疗糖尿病患者的心力衰竭方面显示出临床潜力[179]。

在 STZ 诱导的 DCM 中，Res 全身治疗显著改善心脏功能，心脏肥大和纤维化的减弱、分泌的卷曲相关蛋白 2（secreted frizzled-related protein two gene, sFRP2）和 Wnt/β-catenin 的心脏免疫染色面积百分比降低以及微血管系统的改善相关。用 Res 预处理的体外间充质肝细胞 MSC 细胞活性增强。总之，Res 可以通过减弱 sFRP2 介导的纤维化和下游 Wnt/β-catenin 通路来提高 MSCs 的心脏重塑能力[180]。

在 D-半乳糖诱导的衰老大鼠模型中，Res 降低血压、心脏重量/体重比和心肌细胞大小，上调 CAT 和 SOD1 和 SOD2 的心脏转录水平，降低 MDA 水平，血清抗衰老蛋白 klotho 浓度增加。Res 和维生素 D 的共同给药通过调节心脏的血流动力学参数和抗氧化状态来保护心脏免受衰老引起的损伤。此外，增加血清 klotho 水平可能是 Res 和维生素 D 抗衰老作用的新机制[181]。

ISO 注射诱导的心肌重塑中，Res 减轻心脏功能障碍，包括减轻心脏肥大和心肌细胞纤维化。同时，抑制巨噬细胞 M1 极化，这表现为心脏中 Ly6Clow 巨噬细胞百分比、M1 细胞因子水平和 CD68 表达降低，以及 Ly6Chigh 巨噬细胞百分比、M2 细胞因子水平和 CD206 表达升高。Res 预处理上调 VEGF-B 的表达和 AMPK 的活性，而下调 RAW264.7 细胞和小鼠中磷酸化 NF-κB p65 的表达。因此，Res 在 ISO 诱导的心肌损伤中具有潜在的治疗作用，这可能是通过 VEGF-B/AMPK/NF-κB 通路抑制巨噬细胞的 M1 极化[182]。Res 抑制缺血诱导的心肌衰老信号和 NLRP3 炎性体激活[183]。

Res 和沙库巴曲/缬沙坦单独治疗可显著预防 MI 诱导大鼠的心脏重塑和功能障碍。沙库巴曲/缬沙坦＋白藜芦醇的联合治疗还可以预防心脏异常。单独或联合给药介导的心脏重塑和功能障碍部分归因于减轻心脏氧化应激、炎症和纤维化。研究结果表明，沙库巴曲/缬沙坦和白藜芦醇对减轻心肌梗死的治疗是有益的[184]。

以腹主动脉缩窄（abdominal aortic constriction，AAC）方法在体内建立的压力超负荷诱导的大鼠心肌肥大模型中，Res 治疗改善了压力超负荷大鼠的 LV 功能，并显著减轻了 LV 肥大和心脏纤维化。用 Res 治疗在 AAC 影响下恢复了 Sirt1 和 TGF-β1/p-Smad3 的表达。在新生大鼠心肌细胞中对 Ang Ⅱ 诱导的肥大进行的体外研究证实了 Res 的心脏保护作用。综上所述，Res 激活 Sirt1 可能通过调节 TGF-β1/p-Smad3 信号通路减轻心脏肥大、心脏纤维化并改善心脏功能[185]。

TAC 诱导的小鼠心脏重塑模型中，Res 给药通过抑制 PTEN 降解和下游介质，显著减弱不正常的心脏收缩和舒张功能、纤维化、炎症和氧化应激。Res（100μmol/L）抑制体外 Ang Ⅱ 诱导的成纤维细胞增殖和活化。Res 通过 PTEN/Akt/Smad2/3 和 NF-κB 信号通路减轻压力超负荷诱导的心脏纤维化和舒张功能障碍[186]。

2. 肝

在感染圆腹雅罗鱼体内后囊蚴的叙利亚仓鼠中，驱虫药吡喹酮（praziquantel，PZQ）联合 Res 或线粒体靶向抗氧化剂 SkQ1 显著减轻肝细胞变化，显著减轻胆管细胞增生、胆管增殖、纤维化以及脂滴和糖原颗粒的积累，还观察到 4-羟基壬烯醛的下调。驱虫药物与抗氧化剂 Res 和 SkQ1 的组合可改善宿主氧化应激并减轻 PZQ 对肝实质的不良影响[187]。

在膳食 Res 的尼罗罗非鱼幼鱼肝脏 mRNA 的转录组学结果中，在 15 天、30 天和 45 天分别有 5 个、179 个和 1526 个显著差异表达基因，富集了 8 条与该免疫反应相关的 KEGG 通路，与细胞因子产生、免疫系统、自噬调节、FOXO 信号、类固醇激素生物合成、脂肪酸代谢相关的基因表达显著下降，而与上述途径相关的基因，包括 LEAP-2、PRDX4、MB、HOMER1、MIF、SAT1、CYTBC1-8，以及 ER 中蛋白质加工途径相关的基因 CNE1、TRAM1 显著增加。这些发现表明，Res 激活了一些与免疫和生物过程相关的基因，以增强先天免疫，也提示 Res 高浓度添加或长期给药可能对罗非鱼肝脏产生负面影响[188]。

脂肪干细胞（adipose-derived stem cells，ADSCs）自体移植治疗糖尿病肝功能障碍大鼠模型中，Res 通过 Sirt1 和 IGFIR 表达增加 HG 应激下的 ADSCs 活力。Res 预处理的 ADSCs 在治疗伴有肝功能不全的糖尿病患者的肝功能不全方面显示出潜力[189]。

在代谢综合征（metabolic syndrome，MS）模型中，Res 和槲皮素抑制血清转氨酶活性，减轻肝损伤炎症，其潜在机制与 TLR4、中性粒细胞弹性蛋白酶和嘌呤能受体 P2Y2 相关，进而导致细胞凋亡减少和 LF 减轻，而肝细胞增殖无变化。Res 和槲皮素作为炎症和嘌呤能受体的调节剂可以减轻与 MS 相关的肝损伤[190]。

在一种肝星状细胞模型 GRX 中，Res 降低了 GRX 迁移和 COL-1 收缩，诱导 GRX 培养基中 TNF-α 和 IL-10 含量增加，IL-6 含量降低。Res 不会改变 α-SMA、COL-1 和 GFAP 的蛋白质含量；但 50μmol/L Res 增加了这些激活相关蛋白的含量。总之，这些结果表明 Res 没有降低 GRX 的激活状态，相反，在 50μmol/L 浓度下触发了促激活效应。因此，Res 增加活化标志物并改变 HSCs 炎性细胞因子的释放[191]。

在 CCl_4 诱导的 LF 模型中，Res 和 β-氨基丙腈（β-aminopropionitrile，BAPN）均通过降低胶原纤维束、肝内 HYP 含量和脂氧合酶（lipoxygenase，LOX）蛋白水平显示出抗纤维化作用，并在联合使用时，抗纤维化作用增强。Res 和 BAPN 联合给药可作为靶向 LOX 的一种很有前景的治疗方法[192]。

肝星状细胞细胞系中，Res 显著抑制 HSCS 增殖，降低细胞指数，诱导 HSCs 失活，减少胶原沉积和 α-SMA 表达。此外，Res 还有助于 HSCS 凋亡，上调 Bax 并下调 Bcl-2 表达。Res 增强了 Hippo 通路的激活，并减少了 YES 相关蛋白（yes-associated protein，YAP）和含 PDZ 结合基序的转录共激活因子（transcriptional coactivator with PDZ-binding motif，TAZ）。Res 通过 Hippo 通路抑制 HSCs 活化进而减轻 LF[193]。

在 CCl_4 诱导的肝纤维化模型中，Res 上调了库普弗细胞中 IL-10 和 M（IL-4）标记的表达，包括甘露糖受体 C1 基因（mannose receptor C1，Mrc1）、Mrc2、CD163 和精氨酸酶Ⅰ（arginase Ⅰ，ArgⅠ）；而它略微抑制了 M（LPS）相关因子的水平，包括 iNOS、TNF-α 和 MCP-1。在体外，Res 促进 M（LPS）转换为 M（IL-4）巨噬细胞，并提高了 CD206 的表达。同时，IL-10 在 M（IL-4）和 M（LPS）中均增加。因此，Res 通过产生更多的 IL-10 促进 M（LPS）极化为 M（IL-4）样巨噬细胞来缓解 LF[194]。

在溴氰菊酯诱导肝损伤的鹌鹑模型中，Res 减轻肝脂肪变性、氧化应激、炎症和细胞凋亡。Res 通过激活 Nrf2 表达抑制细胞凋亡诱导氧化应激，然后通过抑制 NF-κB/TNF-α 和 TGF-β1/Smad3 信号通路减轻溴氰菊酯诱导的鹌鹑 LF[195]。

在胆管结扎鼠模型中，Res 减轻肝损伤和纤维化的发展，表现为肝细胞坏死、胆管增生、胶原沉积、肝脏增大、回声增强、结节边界不规则和胆总管扩张。Res 显著降低肝脏 TGF-β1、Smad3、miR-21、TIMP-1、MDA、IL-17a 的水平，同时也减低血液 ALT、AST 和胆红素的水平。30mg/kg 的 Res 可显著防止 BDL 诱导的肝损伤，这与抑制 TGF-β1-Smad3-miR-21 轴以及促纤维化、氧化应激和炎症的生物标志物有关[196]。

在小鼠血吸虫病模型中，Res 治疗通过增加线粒体膜电位和提升 PGC-α

表达（促进线粒体生物发生）来改善线粒体功能，通过减少胶原沉积和纤维化程度来减轻肝损伤，包括减轻肝瘢痕形成。Res还减少了成虫数量并对蠕虫造成了严重的物理损伤。这些结果表明Res上调线粒体生物发生并抑制纤维化[197]。

在CCl_4诱导的肝纤维化大鼠模型和PDGF-BB刺激的HSC-T6细胞模型中，Res可减轻大鼠的LF，增强自噬。Res逆转了miR-20a对PTEN表达的抑制作用，降低了miR-20a的表达，促进了PTEN、PI3K和p-Akt蛋白的表达，从而减轻了LF。总体而言，Res诱导自噬并激活miR-20a介导的PTEN/PI3K/Akt信号通路以减弱LF[198]。

3. 肺

在吸入性颗粒物PM暴露的小鼠模型中，Res治疗消除了PM诱导的肺部炎症和纤维化，并抑制了自噬过程和NLRP3炎性体激活。在体外BEAS-2B细胞中，Res减轻PM2.5诱导的细胞毒性，抑制自噬过程和NLRP3炎性体活性，并减少IL-1β的产生。Res干预通过抑制自噬相关的NLRP3炎性体激活来缓解长期PM暴露导致的肺部炎症和纤维化[199]。在BLM诱导的大鼠肺纤维化模型中，Res通过抑制HIF-1α和NF-κB的表达来减轻BLM诱导的肺纤维化[200]。在使用^{60}Coγ射线源以18 Gy照射小鼠胸部的模型中，Res和α-硫辛酸以及联合给药均可以减轻肺炎和纤维化。尽管Res不能减轻大多数炎症细胞的浸润以及炎症和血管损伤，但α-硫辛酸及其组合能够减轻大多数受损标志物的改变。α-硫辛酸及其与Res的组合能够减轻肺照射后小鼠肺组织中的纤维化，并降低肺炎标志物水平。与药物的单一形式相比，Res和α-硫辛酸的组合具有更有效的缓解作用[201]。

4. 肾

在STZ诱导的糖尿病大鼠肾脏中，Res[15mg/(kg·d)]和雷米普利[10mg/(kg·d)]联合治疗显示早期DN肾小球硬化是可逆的。雷米普利和Res联合治疗可减轻STZ诱导的糖尿病大鼠中由RhoA/Rho相关螺旋卷曲蛋白激酶（Rho-associated coiled coil-forming protein kinase，ROCK）通路调节的早期DN相关肾小球硬化[202]。

在人肾小管上皮细胞系 HK-2 细胞和 UUO 小鼠肾纤维化模型中，低浓度的 Res 通过 Smad3/Smad4 的 Sirt1 依赖性去乙酰化机制降低了 TGF-β 诱导的 EMT。相比之下，长期暴露于高浓度 Res 通过线粒体氧化应激和 ROCK1 介导的无序细胞骨架重塑促进 HK-2 细胞中的 EMT。在体内，低剂量 Res 治疗部分改善了肾功能，而 Res 的高剂量治疗失去了其抗纤维化作用，甚至加重了肾纤维化。然而，UUO 小鼠比正常小鼠更容易受到高 Res 诱导的肾损伤。Res 在肾脏中既能发挥抗纤维化作用，又具有促纤维化作用，因此，在应用时应慎重考虑肾功能受损个体摄入 Res 的风险[203]。

在伴刀豆球蛋白 A（concanavalin A，Con A）诱导的晚期肾小球硬化模型中，Res 预处理可通过上调 Sirtuin1 介导的 Klotho 表达来改善肾小球硬化[204]。

在庆大霉素（gentamicin，GEN）诱导的肾损伤模型中，Res 预处理抑制了 α-SMA、TGF-β1 和 p-Smad2 的表达增加和 E-cadherin 的表达降低。Res 通过抑制氧化应激并可能参与 TGF-β/Smad 信号通路对 GEN 诱导的 EMT 具有保护作用[205]。

在大鼠尿毒症模型中，Res 组 Scr 和 BUN 水平明显下降，肾间质胶原沉积减少，改善了肾组织病变。Res 处理显著提高了 HSP70 和 p-IκBα 的表达，也显著降低了 p-P65 蛋白的水平。总之，Res 通过激活 HSP70 的表达对尿毒症大鼠肾脏起到保护作用[206]。

5. 前列腺

在前列腺纤维化（prostate fibrosis，PF）患者中，接受 Res 治疗的 A 组患者的美国国立卫生研究院慢性前列腺症状指数（National Institutes of Health chronic prostatitis symptom index，NIH-CPSI）和国际前列腺症状评分（international prostate symptom score，IPSS）均显示出显著的症状改善，并且前列腺按摩后前列腺分泌（prostate secretion，EPS）测试的前列腺体积减小和白细胞计数减少。在选定的 PF 患者中使用 Res 治疗 2 个月后发现，Res 在治疗排尿和缓解储存不适症状方面具有药理学优势[207]。

6. 皮肤

在病理性瘢痕成纤维细胞中，Res 抑制 PI3K mRNA 和蛋白质的相对表达量，呈剂量依赖关系。Res 抑制病理性瘢痕成纤维细胞增殖的机制可能与其下调 mTOR 信号通路关键分子 Akt 和 mTOR 的表达有关[208]。在肥厚性瘢痕衍生成纤维细胞中，Res 以剂量依赖性方式显著诱导细胞死亡，并通过下调脑中富含的 Ras 同源蛋白（recombinant human Ras homolog enriched in brain, Rheb）的表达水平来诱导自噬。Res 可能通过 miR-4654/Rheb 轴改变自噬过程。总之，Res 可能通过上调 miR-4654 来促进自噬，进而可能通过直接与 Rheb 的 3′-非翻译区结合来抑制 Rheb 的表达，从而抑制肥厚性瘢痕形成[209]。

在缺氧环境下培养的瘢痕疙瘩来源的成纤维细胞中，Res 可以下调 HIF-1α 和减少胶原合成。Res 能通过靶向 HIF-1α 而抑制瘢痕疙瘩成纤维细胞的增殖并促进其凋亡[210]。

7. 系统硬化

在 BLM 诱导的系统性硬化症（systemic sclerosis，SSc）模型中，Res 靶向基因的富集分析显示 79 条相关通路，其中 27 条参与 SSc。特别是发现 Sirt1/mTOR 信号是关键的调节途径之一。Res 能够提高成纤维细胞中的 Sirt1 水平，并部分逆转 mTOR 依赖性纤维化和炎症的诱导。这些结果表明，Res 是治疗 SSc 可行且有效的选择，mTOR 的治疗靶点可能是治疗 SSc 的潜在替代方案[211]。

8. 眼睛

在人晶状体上皮细胞系 FHL124、人晶状体囊袋模型和中央前上皮作体外细胞模型中，Res 显著阻碍了伤口愈合试验中的细胞迁移。Res 显著抑制 TGF-β2 诱导的 MF 标志物 α-SMA 的表达，并显著抑制 TGF-β2 诱导的基质收缩。因此，Res 可以在两个人类晶状体模型系统中对抗后囊膜混浊相关的生理事件[212]。

9. 鼻

CFTR 基因的突变能导致氯离子转运缺陷，并导致囊性纤维化患者的上呼吸道和下呼吸道慢性细菌感染。Ivacaftor 是一种 CFTR 增强剂。在

G551D CFTR 转染的 Fisher 大鼠甲状腺细胞和含有 CFTR G551D 突变的人鼻窦上皮细胞中，G551D CFTR 介导的氯离子分泌表明 Res 可以增强这些患者的 Ivacaftor 治疗效果并改善囊性纤维化相关的鼻窦炎[213]。

10. 胰腺

在 STZ 诱导糖尿病大鼠的胰腺损伤模型中，ADSCs 移植以及联合 Res 治疗能减轻胰岛缩小、抑制纤维化通路激活、逆转存活信号抑制和凋亡通路表达，同时还能抑制血糖升高。Res 的 ADSCs 预处理增加了生存标志物 p-Akt 的表达，从而提高了 ADSCs 的活力。这项研究表明，Res 的 ADSCs 预处理显示出治疗 T1DM 患者的潜力[214]。

在体外人胰腺星状细胞（human pancreatic stellate cells，hPSCs）中，缺氧通过 HIF-1 诱导 PSC 活化，并且源自活化 PSC 的 IL-6、血管内皮生长因子 A 和基质细胞衍生因子 1 促进胰腺癌的侵袭和上皮-间质转化并抑制细胞凋亡。在胰腺导管腺癌模型小鼠模型中，Res 可抑制缺氧诱导的 PSC 活化，阻断 PSC 与胰腺癌细胞之间的相互作用，并抑制胰腺癌的恶性进展和间质结缔组织增生。总而言之，Res 通过抑制缺氧诱导的 PSCs 活化来改善胰腺癌的恶性进展[215]。

11. 关节

在注射弗氏完全佐剂建立的关节炎-间质性肺病大鼠模型中，Res 治疗显著改善了肺部损伤并防止了促炎细胞因子的产生，抑制 JAK/STAT/NF-κB 受体活化因子（nuclear factor-B receptor activator，RANKL）信号通路。Res 全身治疗通过调节 JAK/STAT/RANKL 信号通路减轻大鼠佐剂性关节炎-间质性肺病[216]。

12. 囊性纤维化

囊性纤维化上皮细胞中，Res 治疗显著恢复了细胞内转运，其机制是恢复微管形成率和微管乙酰化。从机制层面讲，Res 对微管调节和细胞内转运的影响取决于过氧化物酶体增殖物激活受体-γ 信号转导及其作为 HDAC 抑制剂的能力。总之，Res 是一种具有已知抗炎特性的候选化合物，可以恢复囊性纤维化上皮细胞中的微管形成和乙酰化[217]。

13. 神经

在 CCl₄ 诱导的肝硬化模型中，Res 减少血清循环氨、脑液、ALT、AST、TNF-α、IL-1β、4HNE、NF-κB 和 iNOS 水平，同时所有神经元 TJ 蛋白（例如 ZO-1、claudin 5 和 occludin）均降低。Res 可以通过纠正 CCl₄ 诱导的肝硬化小鼠的氨和炎症来恢复神经元紧密连接蛋白[218]。

14. 新剂型

在 HFD 诱导的非酒精性脂肪肝小鼠模型中，A19（一种新型白藜芦醇-姜黄素混合物）显著抑制了脂质积累、肝损伤、肝脏炎症和纤维化。在体外，A19 明显抑制 PA 诱导的炎性细胞因子和纤维化标志物 mRNA 的表达。同时，发现 A19 的抗炎作用与抑制 ERK 信号通路有关。A19 可减轻 HFD 诱导的 NAFLD 疾病[219]。

Vectisol® 是一种新的抗氧化策略，即将 Res 封装在环糊精中，从而提高其生物利用度。在临床前猪肾移植模型中，Vectisol® 静态保存溶液使早期肾小球滤过和近端肾小管功能障碍恢复。延长的随访证实了更高水平的保护，减缓了慢性功能丧失（肌酐血症和蛋白尿）和组织学病变的发生。在机器灌注试验中，使用 Vectisol® 可降低再灌注开始时（去钳位后 30min）的氧化应激和细胞凋亡水平。通过降低循环 SOD 和 ASAT 的早期水平，证实了移植质量的提高。补充剂减缓了慢性功能丧失以及间质纤维化和肾小管萎缩的发生。在保存液中添加 Vectisol® 可显著提高器官保存的性能，并对结果产生长期影响[220]。

在 Ehrlich 腹水瘤小鼠中，Res 及其纳米晶体（NANO-Res）的给药导致腹腔肿瘤细胞增殖显著减少，腹膜血管数量减少，全身毒性低。在组织病理学检查中，与 NANO-Res 相比，Res 观察到更多的肝细胞坏死和凋亡、中央静脉周围的 LF 并伴有轻微脂肪变性。同时，Res 和 NANO-Res 组的几个生化参数略有升高，还观察到伴有近端肾小管坏死和肾小球肿胀的炎症。因此，为了增加与 Res 纳米晶体相关的有益效果并降低风险，还应考虑其他因素，如剂量、遗传因素、健康状况和靶细胞的性质[221]。

第十五节
木犀草素

木犀草素（luteolin，Lut，图 3-22），是一种天然黄酮类化合物，存在于多种植物中。其具有多种药理活性，如消炎、抗过敏、降尿酸、抗肿瘤、抗菌、抗病毒等，临床主要用于止咳、祛痰、消炎、降尿酸以及治疗心血管疾病、"肌萎缩侧索硬化症"和肝炎等。

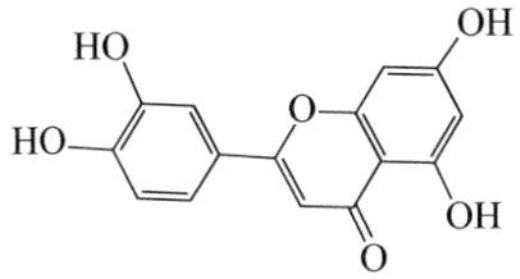

图 3-22　木犀草素分子结构式

1. 心

糖尿病会引发特定的心脏功能和结构异常，从而导致 DCM。在功能上，DCM 的特征是心室功能障碍，但不伴有高血压和缺血性冠状动脉疾病，两个结构特征是 LV 质量和室壁厚度增加以及心肌纤维化。有许多因素共同促进了 DCM 的发生和发展，其中包括高血糖、胰岛素抵抗、脂肪酸代谢增加、血管改变、心肌炎症、重塑和纤维化。越来越多的证据表明，氧化/硝化应激和促炎反应之间的复杂相互作用在启动 DCM 中观察到的细胞异常级联反应中发挥了关键作用。糖尿病和高血糖会使细胞内 ROS 增加，进而诱发损伤和炎症。因此，同时对抗炎症和氧化损伤或许有效地阻止 DCM 的进展。

许多天然存在的化合物具有抗炎和抗氧化活性，这些化合物在 DCM 治疗中具有重要的临床价值。Lut 是一种广泛分布于中草药提取物中的类黄酮，可以预防肺纤维化和肝纤维化。另外，Lut 也可以减少癌细胞的增殖。这些作用被认为是由于氧化应激减轻和炎症反应的减少所致。例如，Lut 通过调节丝裂原活化蛋白激酶通路减轻 LPS 诱导的巨噬细胞损伤和氧化应激。

在相同的实验系统中，炎症反应也被抑制。在心脏组织中，Lut 通过激活抗氧化 Nrf2 信号、恢复线粒体完整性和诱导促生存途径来减轻缺血/再灌注损伤的情况。

Lut 通过抑制炎症和氧化应激减轻纤维化和细胞凋亡来对抗 DCM。其抗炎和抗氧化作用是通过抑制 NF-κB 和诱导 Nrf2 通路介导的。因此，NF-κB 和 Nrf2 在 DCM 发病机制中具有重要作用，Lut 在 DCM 治疗中具有潜在临床应用价值。

2. 肝

LF 的表观遗传调控机制错综复杂，主要包括 DNA 甲基化、组蛋白修饰、microRNA（miRNA）表达等，通过影响纤维化相关基因的表达，HSCs 的活化、增殖、凋亡，以及 MF 分化来调节 LF 的发生和发展。此外，乙酰化和甲基化在 LF 中更常见，并且能够上调基因表达或下调特定基因表达。

肝损伤后，肝脏中主要的胶原蛋白合成细胞 HSCs 被激活并转化为肌成纤维细胞样细胞，促进细胞增殖和存活。HSCs 激活由多种介质驱动，例如趋化因子、活性氧和生长因子。α-SMA 和 TGF-β1 是 HSCs 活化的标志物。Lut 可以降低 α-SMA 和 TGF-β1 的蛋白质表达水平，从而抑制 HSCs 活化并减少胶原蛋白产生[222]。

3. 肺

肺内细胞能合成和分泌多种生长因子，其中细胞因子与细胞因子网络在肺纤维化发展中发挥重要作用。体内外实验证实 TGF-β 能促进肺成纤维细胞增生和胶原蛋白合成，实验性肺纤维化病变中有高水平的 TGF-β，同时伴有 I 型、III 型胶原的 mRNA 表达增高。抗 TGF-β 血清可使纤维化肺组织内 HYP 含量减少，该指标能较好反映出胶原合成水平的高低，减少说明肺组织中胶原含量下降、纤维化程度减轻。Lut 可以抑制 NLRP3 炎症小体的活化，减少 Caspase-1 的激活，抑制炎症介质 IL-1β 和 IL-18 表达，最终降低 TGF-β 的表达，从而起到减轻肺部炎症反应和纤维化的作用。

参考文献

[1] Khodarahmi A, Eshaghian A, Safari F, et al. Quercetin mitigates hepatic insulin resistance in rats with bile duct ligation through modulation of the STAT3/SOCS3/IRS1 signaling pathway[J]. J Food Sci, 2019, 84(10): 3045-3053.

[2] Hohmann M S, Habiel D M, Coelho A L, et al. Quercetin enhances ligand-induced apoptosis in senescent idiopathic pulmonary pibrosis fibroblasts and reduces lung fibrosis *in vivo*[J]. Am J Respir Cell Mol Biol, 2019, 60(1): 28-40.

[3] Takano M, Deguchi J, Senoo S, et al. Suppressive effect of quercetin against bleomycin-induced epithelial-mesenchymal transition in alveolar epithelial cells[J]. Drug Metab Pharmacokinet, 2020, 35(6): 522-526.

[4] Liu T, Yang Q F, Zhang X, et al. Quercetin alleviates kidney fibrosis by reducing renal tubular epithelial cell senescence through the SIRT1/PINK1/mitophagy axis[J]. Life Sci, 2020, 257: 118116.

[5] Liu X H, Sun N, Mo N, et al. Quercetin inhibits kidney fibrosis and the epithelial to mesenchymal transition of the renal tubular system involving suppression of the Sonic Hedgehog signaling pathway[J]. Food Funct, 2019, 10(6): 3782-3797.

[6] Liu Y F, Dai E L, Yang J. Quercetin suppresses glomerulosclerosis and TGF-β signaling in a rat model[J]. Mol Med Rep, 2019, 19(6): 4589-4596.

[7] Rezk A M, Ibrahim I A A E, Mahmoud M F, et al. Quercetin and lithium chloride potentiate the protective effects of carvedilol against renal ischemia-reperfusion injury in high-fructose, high-fat diet-fed Swiss albino mice independent of renal lipid signaling[J]. Chem Biol Interact, 2021, 333:109307.

[8] Xiao Y B, Zhou L, Zhang T T, et al. Anti-fibrosis activity of quercetin attenuates rabbit tracheal stenosis via the TGF-β/AKT/mTOR signaling pathway[J]. Life Sci, 2020, 250: 117552.

[9] Xu J, Tan Y L, Liu Q Y, et al. Quercetin regulates fibrogenic responses of endometrial stromal cell by upregulating miR-145 and inhibiting the TGF-β1/Smad2/Smad3 pathway[J]. Acta Histochem, 2020, 122(7): 151600.

[10] Rahdar A, Hasanein P, Bilal M, et al. Quercetin-loaded F127 nanomicelles: Antioxidant activity and protection against renal injury induced by gentamicin in rats[J]. Life Sci, 2021, 276: 119420.

[11] Zhang Q, Xu D, Guo Q Y, et al. Theranostic quercetin nanoparticle for treatment of hepatic fibrosis[J]. Bioconjug Chem, 2019, 30(11): 2939-2946.

[12] Meng S Y, Yang F, Wang Y Q, et al. Silymarin ameliorates diabetic cardiomyopathy via inhibiting TGF-β1/Smad signaling[J]. Cell Biol Int, 2019, 43(1): 65-72.

[13] Silva C M, Ferrari G D, Alberici L C, et al. Cellular and molecular effects of silymarin on the transdifferentiation processes of LX-2 cells and its connection with lipid metabolism[J]. Mol Cell

Biochem, 2020, 468(1-2): 129-142.

[14] Ali S A, Saifi M A, Godugu C, et al. Silibinin alleviates silica-induced pulmonary fibrosis: Potential role in modulating inflammation and epithelial-mesenchymal transition[J]. Phytother Res, 2021, 35(9): 5290-5304.

[15] Ma Z C, Zang W W, Wang H G, et al. Silibinin enhances anti-renal fibrosis effect of MK-521 via downregulation of TGF-β signaling pathway[J]. Hum Cell, 2020, 33(2): 330-336.

[16] Liu K, Zhou S J, Liu J Y, et al. Silibinin attenuates high-fat diet-induced renal fibrosis of diabetic nephropathy[J]. Drug Des Devel Ther, 2019, 13: 3117-3126.

[17] Liu R G, Wang Q Q, Ding Z Y, et al. Silibinin augments the antifibrotic effect of valsartan through inactivation of TGF-β1 signaling in kidney[J]. Drug Des Devel Ther, 2020, 14: 603-611.

[18] Liu X L, Xu Q, Long X Y, et al. Silibinin-induced autophagy mediated by PPARα-sirt1-AMPK pathway participated in the regulation of type I collagen-enhanced migration in murine 3T3-L1 preadipocytes[J]. Mol Cell Biochem, 2019, 450(1-2): 1-23.

[19] Méndez-Sánchez N, Dibildox-Martinez M, Sosa-Noguera J, et al. Superior silybin bioavailability of silybin-phosphatidylcholine complex in oily-medium soft-gel capsules versus conventional silymarin tablets in healthy volunteers[J]. BMC Pharmacol Toxicol, 2019, 20(1): 5.

[20] Luo J W, Zhang Z W, Zeng Y C, et al. Co-encapsulation of collagenase type I and silibinin in chondroitin sulfate coated multilayered nanoparticles for targeted treatment of liver fibrosis[J]. Carbohydr Polym, 2021, 263: 117964.

[21] Piazzini V, Cinci L, D'Ambrosio M, et al. Solid lipid nanoparticles and chitosan-coated solid lipid nanoparticles as promising tool for silybin delivery: Formulation, characterization, and in vitro evaluation[J]. Curr Drug Deliv, 2019, 16(2): 142-152.

[22] Du Y, Han J B, Zhang H X, et al. Kaempferol prevents against Ang Ⅱ-induced cardiac remodeling through attenuating Ang Ⅱ-induced inflammation and oxidative stress[J]. J Cardiovasc Pharmacol, 2019, 74(4): 326-335.

[23] Xu T F, Huang S, Huang Q R, et al. Kaempferol attenuates liver fibrosis by inhibiting activin receptor-like kinase 5[J]. J Cell Mol Med, 2019, 23(9): 6403-6410.

[24] Liu H Q, Yu H, Cao Z J, et al. Kaempferol modulates autophagy and alleviates silica-induced pulmonary fibrosis[J]. DNA Cell Biol, 2019, 38(12): 1418-1426.

[25] Ji X J, Cao J, Zhang L T, et al. Kaempferol protects renal fibrosis through activating the BMP-7-Smad1/5 signaling pathway[J]. Biol Pharm Bull, 2020, 43(3): 533-539.

[26] Sharma D, Gondaliya P, Tiwari V, et al. Kaempferol attenuates diabetic nephropathy by inhibiting RhoA/Rho-kinase mediated inflammatory signalling[J]. Biomed Pharmacother, 2019, 109: 1610-1619.

[27] Sekiguchi A, Motegi S I, Fujiwara C, et al. Inhibitory effect of kaempferol on skin fibrosis in

systemic sclerosis by the suppression of oxidative stress[J]. J Dermatol Sci, 2019, 96(1): 8-17.

[28] Özay Y, Güzel S, Yumrutaş Ö, et al. Wound healing effect of kaempferol in diabetic and nondiabetic rats[J]. J Surg Res, 2019, 233: 284-296.

[29] Wang F, Fan K, Zhao Y, et al. Apigenin attenuates TGF-β1-stimulated cardiac fibroblast differentiation and extracellular matrix production by targeting miR-155-5p/c-Ski/Smad pathway [J]. J Ethnopharmacol, 2021, 265: 113195.

[30] Feng W, Ying Z, Ke F, et al. Apigenin suppresses TGF-β1-induced cardiac fibroblast differentiation and collagen synthesis through the downregulation of HIF-1α expression by miR-122-5p[J]. Phytomedicine, 2021, 83: 153481.

[31] Qiao M, Yang J H, Zhu Y, et al. Transcriptomics and proteomics analysis of system-level mechanisms in the liver of apigenin-treated fibrotic rats[J]. Life Sci, 2020, 248: 117475.

[32] Zheng S H, Cao P C, Yin Z Q, et al. Apigenin protects mice against 3,5-diethoxycarbonyl-1,4-dihydrocollidine-induced cholestasis[J]. Food Funct, 2021, 12(5): 2323-2334.

[33] Ogura Y, Kitada M, Xu J, et al. CD38 inhibition by apigenin ameliorates mitochondrial oxidative stress through restoration of the intracellular $NAD^+/NADH$ ratio and Sirt3 activity in renal tubular cells in diabetic rats[J]. Aging (Albany NY), 2020, 12(12): 11325-11336.

[34] Li N N, Wang Z, Sun T, et al. Apigenin alleviates renal fibroblast activation through AMPK and ERK signaling pathways in vitro[J]. Curr Pharm Biotechnol, 2020, 21(11): 1107-1118.

[35] Jiao R, Chen H, Wan Q, et al. Apigenin inhibits fibroblast proliferation and reduces epidural fibrosis by regulating Wnt3a/β-catenin signaling pathway[J]. J Orthop Surg Res, 2019, 14(1): 258.

[36] Zhang L P, Wang S D, Li Y Y, et al. Cardioprotective effect of icariin against myocardial fibrosis and its molecular mechanism in diabetic cardiomyopathy based on network pharmacology: Role of ICA in DCM[J]. Phytomedicine, 2021, 91: 153607.

[37] Qi C L, Shao Y Z, Liu X S, et al. The cardioprotective effects of icariin on the isoprenaline-induced takotsubo-like rat model: Involvement of reactive oxygen species and the TLR4/NF-κB signaling pathway[J]. Int Immunopharmacol, 2019, 74: 105733.

[38] Ye L, Yu Y P, Zhao Y P. Icariin-induced miR-875-5p attenuates epithelial-mesenchymal transition by targeting hedgehog signaling in liver fibrosis[J]. J Gastroenterol Hepatol, 2020, 35(3): 482-491.

[39] Chen H A, Chen C M, Guan S S, et al. The antifibrotic and anti-inflammatory effects of icariin on the kidney in a unilateral ureteral obstruction mouse model [J]. Phytomedicine, 2019, 59: 152917.

[40] Wei Y C, Wu Y, Feng K, et al. Astragaloside Ⅳ inhibits cardiac fibrosis via miR-135a-TRPM7-TGF-β/Smads pathway[J]. J Ethnopharmacol, 2020, 249: 112404.

[41] Zhu Y B, Qian X, Li J J, et al. Astragaloside Ⅳ protects H9C2(2-1) cardiomyocytes from high glucose-induced injury via miR-34a-mediated autophagy pathway[J]. Artif Cells Nanomed Biotechnol, 2019, 47(1): 4172-4181.

[42] Cheng S Y, Zhang X X, Feng Q, et al. Astragaloside Ⅳ exerts angiogenesis and cardioprotection after myocardial infarction via regulating PTEN/PI3K/Akt signaling pathway[J]. Life Sci, 2019, 227: 82-93.

[43] Jiang S, Jiao G Y, Chen Y Q, et al. Astragaloside Ⅳ attenuates chronic intermittent hypoxia-induced myocardial injury by modulating Ca^{2+} homeostasis[J]. Cell Biochem Funct, 2020, 38(6): 710-720.

[44] Lin J M, Fang L J, Li H, et al. Astragaloside Ⅳ alleviates doxorubicin induced cardiomyopathy by inhibiting NADPH oxidase derived oxidative stress[J]. Eur J Pharmacol, 2019, 859: 172490.

[45] Luo L F, Guan P, Qin L Y, et al. Astragaloside Ⅳ inhibits adriamycin-induced cardiac ferroptosis by enhancing Nrf2 signaling[J]. Mol Cell Biochem, 2021, 476(7): 2603-2611.

[46] Li N N, Wu K, Feng F F, et al. Astragaloside Ⅳ alleviates silica-induced pulmonary fibrosis via inactivation of the TGF-β1/Smad2/3 signaling pathway[J]. Int J Mol Med, 2021, 47(3): 16.

[47] Li N N, Feng F F, Wu K, et al. Inhibitory effects of astragaloside Ⅳ on silica-induced pulmonary fibrosis via inactivating TGF-β1/Smad3 signaling[J]. Biomed Pharmacother, 2019, 119: 109387.

[48] Wang X L, Gao Y B, Tian N X, et al. Astragaloside Ⅳ inhibits glucose-induced epithelial-mesenchymal transition of podocytes through autophagy enhancement via the SIRT-NF-κB p65 axis[J]. Sci Rep, 2019, 9(1): 323.

[49] Su Y, Chen Q Q, Ma K K, et al. Astragaloside Ⅳ inhibits palmitate-mediated oxidative stress and fibrosis in human glomerular mesangial cells via downregulation of CD36 expression[J]. Pharmacol Rep, 2019, 71(2): 319-329.

[50] Zhang Y D, Tao C H, Xuan C, et al. Transcriptomic analysis reveals the protection of astragaloside Ⅳ against diabetic nephropathy by modulating inflammation[J]. Oxid Med Cell Longev, 2020, 9542165.

[51] Chen X, Yang Y, Liu C X, et al. Astragaloside Ⅳ ameliorates high glucose-induced renal tubular epithelial-mesenchymal transition by blocking mTORC1/p70S6K signaling in HK-2 cells[J]. Int J Mol Med, 2019, 43(2): 709-716.

[52] Zhu W W, Zhang X, Gao K, et al. Effect of astragaloside Ⅳ and the role of nuclear receptor RXRα in human peritoneal mesothelial cells in high glucose-based peritoneal dialysis fluids[J]. Mol Med Rep, 2019, 20(4): 3829-3839.

[53] Kasetti R B, Maddineni P, Kodati B, et al. Astragaloside Ⅳ attenuates ocular hypertension in a mouse model of TGFβ2 induced primary open angle glaucoma[J]. Int J Mol Sci, 2021, 22(22): 12508.

[54] Ye S J, Luo W, Khan Z A, et al. Celastrol attenuates angiotensin Ⅱ-induced cardiac remodeling by targeting STAT3[J]. Circ Res, 2020, 126(8): 1007-1023.

[55] Wang Y Q, Li C L, Gu J Y, et al. Celastrol exerts anti-inflammatory effect in liver fibrosis via activation of AMPK-SIRT3 signalling[J]. J Cell Mol Med, 2020, 24(1): 941-953.

[56] Kurosawa R, Satoh K, Nakata T, et al. Identification of celastrol as a novel therapeutic agent for pulmonary arterial hypertension and right ventricular failure through suppression of Bsg (Basigin)/CyPA (Cyclophilin A)[J]. Arterioscler Thromb Vasc Biol, 2021, 41(3): 1205-1217.

[57] Li R, Li Y, Zhang J, et al. Targeted delivery of celastrol to renal interstitial myofibroblasts using fibronectin-binding liposomes attenuates renal fibrosis and reduces systemic toxicity[J]. J Control Release, 2020, 320: 32-44.

[58] Jiang X W, Chen S, Zhang Q S, et al. Celastrol is a novel selective agonist of cannabinoid receptor 2 with anti-inflammatory and anti-fibrotic activity in a mouse model of systemic sclerosis[J]. Phytomedicine, 2020, 67: 153160.

[59] Lu X H, Gong S, Wang X J, et al. Celastrol exerts cardioprotective effect in rheumatoid arthritis by inhibiting TLR2/HMGB1 signaling pathway-mediated autophagy[J]. Int Arch Allergy Immunol, 2021, 182(12): 1245-1254.

[60] Jiang L, Sun J J, Wang P. Tripterine emerges as a potential anti-scarring agent in NIH/3T3 cells by repressing ANRIL[J]. Gen Physiol Biophys, 2020, 39(4): 355-362.

[61] Pan X C, Liu Y, Cen Y Y, et al. Dual role of triptolide in interrupting the NLRP3 inflammasome pathway to attenuate cardiac fibrosis[J]. Int J Mol Sci, 2019, 20(2): 360.

[62] Li W, Gong K Z, Ding Y, et al. Effects of triptolide and methotrexate nanosuspensions on left ventricular remodeling in autoimmune myocarditis rats[J]. Int J Nanomedicine, 2019, 14: 851-863.

[63] Huang R S, Guo F, Li Y P, et al. Activation of AMPK by triptolide alleviates nonalcoholic fatty liver disease by improving hepatic lipid metabolism, inflammation and fibrosis[J]. Phytomedicine, 2021, 92: 153739.

[64] Guo K N, Chen J R, Chen Z J, et al. Triptolide alleviates radiation-induced pulmonary fibrosis via inhibiting IKKβ stimulated LOX production[J]. Biochem Biophys Res Commun, 2020, 527(1): 283-288.

[65] Zhang J, Zhu M J, Zhang S Y, et al. Triptolide attenuates renal damage by limiting inflammatory responses in DOCA-salt hypertension[J]. Int Immunopharmacol, 2020, 89(Pt A): 107035.

[66] Dai J H, Sun Y, Chen D Y, et al. Negative regulation of PI3K/AKT/mTOR axis regulates fibroblast proliferation, apoptosis and autophagy play a vital role in triptolide-induced epidural fibrosis reduction[J]. Eur J Pharmacol, 2019, 864: 172724.

[67] Chen M, Wang J M, Wang D, et al. Triptolide inhibits migration and proliferation of fibroblasts

from ileocolonic anastomosis of patients with Crohn's disease via regulating the miR-16-1/HSP70 pathway[J]. Mol Med Rep, 2019, 19(6): 4841-4851.

[68] Shen S, Luo J T, Ye J P. Artesunate alleviates schistosomiasis-induced liver fibrosis by downregulation of mitochondrial complex Ⅰ subunit NDUFB8 and complex Ⅲ subunit UQCRC2 in hepatic stellate cells[J]. Acta Trop, 2021, 214: 105781.

[69] Wan Q, Chen H, Li X L, et al. Artesunate inhibits fibroblasts proliferation and reduces surgery-induced epidural fibrosis via the autophagy-mediated p53/p21waf1/cip1 pathway[J]. Eur J Pharmacol, 2019, 842: 197-207.

[70] Wan Q, Chen H, Xiong G R, et al. Artesunate protects against surgery-induced knee arthrofibrosis by activatingbeclin-1-mediated autophagy via inhibition of mTOR signaling[J]. Eur J Pharmacol, 2019, 854: 149-158.

[71] Chen H, Tao J, Wang J C, et al. Artesunate prevents knee intraarticular adhesion via PRKR-like ER kinase (PERK) signal pathway[J]. J Orthop Surg Res, 2019, 14(1): 448.

[72] Nong X L, Rajbanshi G, Chen L, et al. Effect of artesunate and relation with TGF-β1 and SMAD3 signaling on experimental hypertrophic scar model in rabbit ear[J]. Arch Dermatol Res, 2019, 311(10): 761-772.

[73] Larson S A, Dolivo D M, Dominko T. Artesunate inhibits myofibroblast formation via induction of apoptosis and antagonism of pro-fibrotic gene expression in human dermal fibroblasts[J]. Cell Biol Int, 2019, 43(11): 1317-1322.

[74] Ma L, Wang X D, Li W, et al. Conjugation of ginsenoside with dietary amino acids: A promising strategy to suppress cell proliferation and induce apoptosis in activated hepatic stellate cells[J]. J Agric Food Chem, 2019, 67(36): 10245-10255.

[75] Li X, Cui X, Zhou S, et al. The novel ginsenoside AD2 prevents angiotensin Ⅱ-induced connexin 40 and connexin 43 dysregulation by activating AMP kinase signaling in perfused beating rat atria [J]. Chem Biol Interact, 2021, 339: 109430.

[76] Su G Y, Li Z Y, Wang R, et al. Signaling pathways involved in p38-ERK and inflammatory factors mediated the anti-fibrosis effect of AD-2 on thioacetamide-induced liver injury in mice[J]. Food Funct, 2019, 10(7): 3992-4000.

[77] Jiang L J, Yin X J, Chen Y H, et al. Proteomic analysis reveals ginsenoside Rb1 attenuates myocardial ischemia/reperfusion injury through inhibiting ROS production from mitochondrial complex Ⅰ[J]. Theranostics, 2021, 11(4): 1703-1720.

[78] Liu X H, Chen J W, Sun N, et al. Ginsenoside Rb1 ameliorates autophagy via the AMPK/mTOR pathway in renal tubular epithelial cells *in vitro* and *in vivo*[J]. Int J Biol Macromol, 2020, 163: 996-1009.

[79] Ke S Y, Liu D H, Wu L, et al. Ginsenoside Rb1 ameliorates age-related myocardial dysfunction

by regulating the NF-κB signaling pathway[J]. Am J Chin Med, 2020, 48(6): 1369-1383.

[80] Zhang Y, Ji H X, Qiao O, et al. Nanoparticle conjugation of ginsenoside Rb3 inhibits myocardial fibrosis by regulating PPARα pathway[J]. Biomed Pharmacother, 2021, 139: 111630.

[81] Zhang N N, An X B, Lang P P, et al. Ginsenoside Rdcontributes the attenuation of cardiac hypertrophy *in vivo* and *in vitro*[J]. Biomed Pharmacother, 2019, 109: 1016-1023.

[82] Yu Y H, Sun J H, Liu J G, et al. Ginsenoside Re preserves cardiac function and ameliorates left ventricular remodeling in a rat model of myocardial infarction[J]. J Cardiovasc Pharmacol, 2020, 75(1): 91-97.

[83] Mo C, Xie S W, Zeng T, et al. Ginsenoside-Rg1 acts as an IDO1 inhibitor, protects against liver fibrosis via alleviating IDO1-mediatedthe inhibition of DCs maturation[J]. Phytomedicine, 2021, 84: 153524.

[84] Li X H, Xiang N, Wang Z R. Ginsenoside Rg2 attenuates myocardial fibrosis and improves cardiac function after myocardial infarctionvia AKT signaling pathway[J]. Biosci Biotechnol Biochem, 2020, 84(11): 2199-2206.

[85] Wang Q W, Fu W W, Yu X F, et al. Ginsenoside Rg2 alleviates myocardial fibrosis by regulating TGF-β1/Smad signalling pathway[J]. Pharm Biol, 2021, 59(1): 106-113.

[86] Liu X X, Mi X J, Wang Z, et al. Ginsenoside Rg3 promotes regression from hepatic fibrosis through reducing inflammation-mediated autophagy signaling pathway[J]. Cell Death Dis, 2020, 11(6): 454.

[87] Li L, Wang Y L, Guo R, et al. Ginsenoside Rg3-loaded, reactive oxygen species-responsive polymeric nanoparticles for alleviating myocardial ischemia-reperfusion injury[J]. J Control Release, 2020, 317: 259-272.

[88] Yang L, Chen P P, Luo M, et al. Inhibitory effects of total ginsenoside on bleomycin-induced pulmonary fibrosis in mice[J]. Biomed Pharmacother, 2019, 114: 108851.

[89] Fu Z, Xu Y S, Cai C Q. Ginsenoside Rg3 inhibits pulmonary fibrosis by preventing HIF-1α nuclear localisation[J]. BMC Pulm Med, 2021, 21(1): 70.

[90] Zhou T, Sun L, Yang S, et al. 20(S)-Ginsenoside Rg3 protects kidney from diabetic kidney disease via renal inflammation depression in diabetic rats[J]. J Diabetes Res, 2020, 2020: 7152176.

[91] Yan X, Zhang W, Kong F W, et al. Ginsenoside Rg3 reduces epithelial-mesenchymal transition induced by transforming growth factor-β1 by inactivation of AKT in HMrSV5 peritoneal mesothelial cells[J]. Med Sci Monit, 2019, 25: 6972-6979.

[92] Qu L L, Ma X X, Fan D D. Ginsenoside Rk3 suppresses hepatocellular carcinoma development through targeting the gut-liver axis[J]. J Agric Food Chem, 2021, 69(35): 10121-10137.

[93] Song J, Cui Z Y, Lian L H, et al. 20S-Protopanaxatriol ameliorates hepatic fibrosis, potentially

involving FXR-mediated inflammatory signaling cascades[J]. J Agric Food Chem, 2020, 68(31): 8195-8204.

[94] Wang F, Park J S, Ma Y Q, et al. Ginseng saponin enriched in Rh1 and Rg2 ameliorates nonalcoholic fatty liver disease by inhibiting inflammasome activation[J]. Nutrients, 2021, 13(3): 856.

[95] Yang N N, Chen H Y, Gao Y, et al. Tanshinone Ⅱ A exerts therapeutic effects by acting on endogenous stem cells in rats with liver cirrhosis[J]. Biomed Pharmacother, 2020, 132: 110815.

[96] Feng F F, Cheng P, Zhang H N, et al. The protective role of tanshinone Ⅱ A in silicosis rat model via TGF-β1/Smad signaling suppression, NOX4 inhibition and Nrf2/ARE signaling activation[J]. Drug Des Devel Ther, 2019, 13: 4275-4290.

[97] Feng F F, Li N N, Cheng P, et al. Tanshinone Ⅱ A attenuates silica-induced pulmonary fibrosis via inhibitionof TGF-β1-Smad signaling pathway[J]. Biomed Pharmacother, 2020, 121: 109586.

[98] Feng F F, Cheng P, Xu S H, et al. Tanshinone Ⅱ A attenuates silica-induced pulmonary fibrosis via Nrf2-mediated inhibition of EMT and TGF-β1/Smad signaling[J]. Chem Biol Interact, 2020, 319: 109024.

[99] Xu S J, He L J, Ding K K, et al. Tanshinone Ⅱ A ameliorates streptozotocin-induced diabetic nephropathy, partly by attenuating PERK pathway-induced fibrosis[J]. Drug Des Devel Ther, 2020, 14: 5773-5782.

[100] Jiang Y, Hu F F, Li Q, et al. Tanshinone Ⅱ A ameliorates the bleomycin-induced endothelial-to-mesenchymal transition via the Akt/mTOR/p70S6K pathway in a murine model of systemic sclerosis[J]. Int Immunopharmacol, 2019, 77: 105968.

[101] Zhou H Y, Jiang S, Li P F, et al. Improved tendon healing by a combination of tanshinone Ⅱ A and miR-29b inhibitor treatment through preventing tendon adhesion and enhancing tendon strength[J]. Int J Med Sci, 2020, 17(8): 1083-1094.

[102] Hsiao Y W, Tsai Y N, Huang Y T, et al. Rhodiola crenulata reduces ventricular arrhythmia through mitigating the activation of IL-17 and inhibiting the MAPK signaling pathway[J]. Cardiovasc Drugs Ther, 2021, 35(5): 889-900.

[103] Ni J, Li Y M, Xu Y W, et al. Salidroside protects against cardiomyocyte apoptosis and ventricular remodeling by AKT/HO-1 signaling pathways in a diabetic cardiomyopathy mouse model[J]. Phytomedicine, 2021, 82: 153406.

[104] Chen P S, Liu J, Ruan H Y, et al. Protective effects of salidroside on cardiac function in mice with myocardial infarction[J]. Sci Rep, 2019, 9(1): 18127.

[105] Hu M L, Zhang D R, Xu H Y, et al. Salidroside activates the AMP-activated protein kinase pathway to suppress nonalcoholic steatohepatitis in mice[J]. Hepatology, 2021, 74(6): 3056-3073.

［106］Li R，Guo Y J，Zhang Y M，et al. Salidroside ameliorates renal interstitial fibrosis by inhibiting the TLR4/NF-κB and MAPK signaling pathways［J］. Int J Mol Sci，2019，20(5)：1103.

［107］Xue H Y，Li P P，Luo Y S，et al. Salidroside stimulates the Sirt1/PGC-1α axis and ameliorates diabetic nephropathy in mice［J］. Phytomedicine，2019，54：240-247.

［108］Huang X Z，Xue H Y，Ma J Y，et al. Salidroside ameliorates adriamycin nephropathy in mice by inhibiting β-catenin activity［J］. J Cell Mol Med，2019，23(6)：4443-4453.

［109］Huang Y P，Han X D，Tang J Y，et al. Salidroside inhibits endothelial-mesenchymal transition via the KLF4/eNOS signaling pathway［J］. Mol Med Rep，2021，24(4)：692.

［110］Gao H，Peng L，Li C，et al. Salidroside alleviates cartilage degenerationthrough NF-κB pathway in osteoarthritis rats［J］. Drug Des Devel Ther，2020，14：1445-1454.

［111］Ye M Y，Zhao F，Ma K，et al. Enhanced effects of salidroside on erectile function and corpora cavernosa autophagy in a cavernous nerve injury rat model［J］. Andrologia，2021，53(6)：e14044.

［112］Wu X，Huang L T，Zhou X L，et al. Curcumin protects cardiomyopathy damage through inhibiting the production of reactive oxygen species in type 2 diabetic mice［J］. Biochem Biophys Res Commun，2020，530(1)：15-21.

［113］Gbr A A，Abdel Baky N A，Mohamed E A，et al. Cardioprotective effect of pioglitazone and curcumin against diabetic cardiomyopathy in type 1 diabetes mellitus：impact on CaMKII/NF-κB/TGF-β1 and PPAR-γ signaling pathway［J］. Naunyn Schmiedebergs Arch Pharmacol，2021，394(2)：349-360.

［114］Yue H H，Zhao X S，Liang W T，et al. Curcumin，novel application in reversing myocardial fibrosis in the treatment for atrial fibrillation from the perspective of transcriptomics in rat model ［J］. Biomed Pharmacother，2022，146：112522.

［115］Sunagawa Y，Funamoto M，Shimizu K，et al. Curcumin，an inhibitor of p300-HAT activity，suppresses the development of hypertension-induced left ventricular hypertrophy with preserved ejection fraction in Dahl rats［J］. Nutrients，2021，13(8)：2608.

［116］Han X Q，Xu S Q，Lin J G. Curcumin recovers intracellular lipid droplet formation through increasing Perilipin 5 gene expression in activated hepatic stellate cells *in vitro*［J］. Curr Med Sci，2019，39(5)：766-777.

［117］Hernández-Aquino E，Quezada-Ramírez M A，Silva-Olivares A，et al. Curcumin downregulates Smad pathways and reduces hepatic stellate cells activation in experimental fibrosis［J］. Ann Hepatol，2020，19(5)：497-506.

［118］Nozari E，Moradi A，Samadi M. Effect of atorvastatin，curcumin，and quercetin on miR-21 and miR-122 and their correlation with TGFβ1 expression in experimental liver fibrosis［J］. Life Sci，2020，259：118293.

[119] Gowifel A M H, Khalil M G, Nada S A, et al. Combination of pomegranate extract and curcumin ameliorates thioacetamide-induced liver fibrosis in rats: impact on TGF-β/Smad3 and NF-κB signaling pathways[J]. Toxicol Mech Methods, 2020, 30(8): 620-633.

[120] Saadati S, Hatami B, Yari Z, et al. The effects of curcumin supplementation on liver enzymes, lipid profile, glucose homeostasis, and hepatic steatosis and fibrosis in patients with non-alcoholic fatty liver disease[J]. Eur J Clin Nutr, 2019, 73(3): 441-449.

[121] Abo-Zaid M A, Shaheen E S, Ismail A H. Immunomodulatory effect of curcumin on hepatic cirrhosis in experimental rats[J]. J Food Biochem, 2020, 44(6): e13219.

[122] Lu S, Zhao H S, Zhou Y J, et al. Curcumin affects leptin-induced expression of methionine adenosyltransferase 2A in hepatic stellate cells by inhibition of JNK signaling[J]. Pharmacology, 2021, 106(7-8): 426-434.

[123] Nouri-Vaskeh M, Malek Mahdavi A, Afshan H, et al. Effect of curcumin supplementation on disease severity in patients with liver cirrhosis: A randomized controlled trial[J]. Phytother Res, 2020, 34(6): 1446-1454.

[124] Nouri-Vaskeh M, Afshan H, Malek Mahdavi A, et al. Curcumin ameliorates health-related quality of life in patients with liver cirrhosis: A randomized, double-blind placebo-controlled trial [J]. Complement Ther Med, 2020, 49: 102351.

[125] Fatima Zaidi S N, Mahboob T. Hepatoprotective role of curcumin in rat liver cirrhosis[J]. Pak J Pharm Sci, 2020, 33(4): 1519-1525.

[126] Khodarahmi A, Javidmehr D, Eshaghian A, et al. Curcumin exerts hepatoprotection via overexpression of paraoxonase-1 and its regulatory genes in ratsundergone bile duct ligation[J]. J Basic Clin Physiol Pharmacol, 2020, 32(5): 969-977.

[127] Hu X, Zhou Y J. Curcumin reduces methionine adenosyltransferase 2B expression by interrupting phosphorylation of p38 MAPK in hepatic stellate cells[J]. Eur J Pharmacol, 2020, 886: 173424.

[128] Elswefy S E, Abdallah F R, Wahba A S, et al. Antifibrotic effect of curcumin, N-acetyl cysteine and propolis extract against bisphenol A-induced hepatotoxicity in rats: Prophylaxis versus co-treatment[J]. Life Sci, 2020, 260: 118245.

[129] Macías-Pérez J R, Vázquez-López B J, Muñoz-Ortega M H, et al. Curcumin and α/β-adrenergic antagonists cotreatment reverse liver cirrhosis in hamsters: Participation of Nrf-2 and NF-κB[J]. J Immunol Res, 2019, 2019: 3019794.

[130] Ibrahim K G, Chivandi E, Nkomozepi P, et al. The long-term protective effects of neonatal administration of curcumin against nonalcoholic steatohepatitis in high-fructose-fed adolescent rats[J]. Physiol Rep, 2019, 7(6): e14032.

[131] Abd El-Hameed N M, Abd El-Aleem S A, Khattab M A, et al. Curcumin activation of nuclear

factor E2-related factor 2 gene (Nrf2): Prophylactic and therapeutic effect in nonalcoholic steatohepatitis (NASH)[J]. Life Sci, 2021, 285: 119983.

[132] Mirhafez S R, Azimi-Nezhad M, Dehabeh M, et al. The effect of curcumin phytosome on the treatment of patients with non-alcoholic fatty liver disease: A double-blind, randomized, placebo-controlled trial[J]. Adv Exp Med Biol, 2021, 1308: 25-35.

[133] Kong D S, Zhang Z L, Chen L P, et al. Curcumin blunts epithelial-mesenchymal transition of hepatocytes to alleviate hepatic fibrosis through regulating oxidative stress and autophagy[J]. Redox Biol, 2020, 36: 101600.

[134] Shaikh S B, Najar M A, Prabhu A, et al. The unique molecular targets associated antioxidant and antifibrotic activity of curcumin in *in vitro* model of acute lung injury: A proteomic approach [J]. Biofactors, 2021, 47(4): 627-644.

[135] Gouda M M, Rex D A B, Es S P, et al. Proteomics analysis revealed the importance of inflammation-mediated downstream pathways and the protective role of curcumin in bleomycin-induced pulmonary fibrosis in C57BL/6 mice[J]. J Proteome Res, 2020, 19(8): 2950-2963.

[136] Shaikh S B, Prabhu A, Bhandary Y P. Curcumin suppresses epithelial growth factor receptor (EGFR) and proliferative protein (Ki 67) in acute lung injury and lung fibrosis *in vitro* and *in vivo*[J]. Endocr Metab Immune Disord Drug Targets, 2020, 20(4): 558-563.

[137] Durairaj P, Venkatesan S, Narayanan V, et al. Protective effects of curcumin on bleomycin-induced changes in lung glycoproteins[J]. Mol Cell Biochem, 2020, 469(1-2): 159-167.

[138] Sun C B, Ying Y, Wu Q Y, et al. The main active components of *Curcuma zedoaria* reduces collagen deposition in human lung fibroblast via autophagy[J]. Mol Immunol, 2020, 124: 109-116.

[139] Shaikh S B, Prabhakar Bhandary Y. Effect of curcumin on IL-17A mediated pulmonary AMPK kinase/cyclooxygenase-2 expressions via activation of NFκB in bleomycin-induced acute lung injury*in vivo*[J]. Int Immunopharmacol, 2020, 85: 106676.

[140] Durairaj P, Venkatesan S, Narayanan V, et al. Curcumin inhibition of bleomycin-induced changes in lung collagen synthesis, deposition and assembly[J]. Mol Biol Rep, 2021, 48(12): 7775-7785.

[141] Ke S W, Zhang Y B, Lan Z H, et al. Curcumin protects murine lung mesenchymal stem cells from H_2O_2 by modulating the Akt/Nrf2/HO-1 pathway[J]. J Int Med Res, 2020, 48(4): 1.

[142] Tyagi N, Singh D K, Dash D, et al. Curcumin modulates paraquat-induced epithelial to mesenchymal transition by regulating transforming growth factor-β (TGF-β) in A549 cells[J]. Inflammation, 2019, 42(4): 1441-1455.

[143] Saidi A, Kasabova M, Vanderlynden L, et al. Curcumin inhibits the TGF-β1-dependent differentiation of lung fibroblasts via PPARγ-driven upregulation of cathepsins B and L[J]. Sci

Rep, 2019, 9(1): 491.

[144] Chang W A, Chen C M, Sheu C C, et al. The potential effects of curcumin on pulmonary fibroblasts of idiopathic pulmonary fibrosis (IPF)-approaching with next-generation sequencing and bioinformatics[J]. Molecules, 2020, 25(22): 5458.

[145] He Y B, Lang X J, Cheng D, et al. Curcumin ameliorates chronic renal failure in 5/6 nephrectomized rats by regulation of the mTOR/HIF-1α/VEGF signaling pathway[J]. Biol Pharm Bull, 2019, 42(6): 886-891.

[146] Lu M M, Li H, Liu W L, et al. Curcumin attenuates renal interstitial fibrosis by regulating autophagy and retaining mitochondrial function in unilateral ureteral obstruction rats[J]. Basic Clin Pharmacol Toxicol, 2021, 128(4): 594-604.

[147] Maghmomeh A O, El-Gayar A M, El-Karef A, et al. Arsenic trioxide and curcumin attenuate cisplatin-induced renal fibrosis in rats through targeting Hedgehog signaling[J]. Naunyn Schmiedebergs Arch Pharmacol, 2020, 393(3): 303-313.

[148] Chen F, Xie Y, Lv Q, et al. Curcumin mediates repulsive guidance molecule B (RGMb) in the treatment mechanism of renal fibrosis induced by unilateral ureteral obstruction[J]. Ren Fail, 2021, 43(1): 1496-1505.

[149] Xu X L, Wang H F, Guo D D, et al. Curcumin modulates gut microbiota and improves renal function in rats with uric acid nephropathy[J]. Ren Fail, 2021, 43(1): 1063-1075.

[150] Li Y H, Zhang J, Liu H Y, et al. Curcumin ameliorates glyoxylate-induced calcium oxalate deposition and renal injuries in mice[J]. Phytomedicine, 2019, 61: 152861.

[151] Chen X, Chen X L, Shi X X, et al. Curcumin attenuates endothelial cell fibrosis through inhibiting endothelial-interstitial transformation[J]. Clin Exp Pharmacol Physiol, 2020, 47(7): 1182-1192.

[152] Chandrashekar A, Annigeri R G, Va U, et al. A clinicobiochemical evaluation of curcumin as gel and as buccal mucoadhesive patches in the management of oral submucous fibrosis[J]. Oral Surg Oral Med Oral Pathol Oral Radiol, 2021, 131(4): 428-434.

[153] Rai A, Kaur M, Gombra V, et al. Comparative evaluation of curcumin and antioxidants in the management of oral submucous fibrosis[J]. J Investig Clin Dent, 2019, 10(4): e12464.

[154] Lanjekar A B, Bhowate R R, Bakhle S, et al. Comparison of efficacy of topical curcumin gel with triamcinolone-hyaluronidase gel individually and in combination in the treatment of oral submucous fibrosis[J]. J Contemp Dent Pract, 2020, 21(1): 83-90.

[155] Piyush P, Mahajan A, Singh K, et al. Comparison of therapeutic response of lycopene and curcumin in oral submucous fibrosis: A randomized controlled trial[J]. Oral Dis, 2019, 25(1): 73-79.

[156] Nerkar Rajbhoj A, Kulkarni T M, Shete A, et al. A comparative study toevaluate efficacy of

curcumin and aloe vera gel along with oral physiotherapy in the management of oral submucous fibrosis: A randomized clinical trial[J]. Asian Pac J Cancer Prev, 2021, 22(S1): 107-112.

[157] Ismailoglu O, Kizilay Z, Cetin N K, et al. Effect of curcumin on the formation of epidural fibrosis in an experimental laminectomy model in rats[J]. Turk Neurosurg, 2019, 29(3): 440-444.

[158] Demirel C, Turkoz D, Yazicioglu I M, et al. The preventive effect of curcumin on the experimental rat epidural fibrosis model[J]. World Neurosurg, 2021, 145: e141-e148.

[159] Zhao J L, Zhang T, Shao X, et al. Curcumin ameliorates peritoneal fibrosis via inhibition of transforming growth factor-activated kinase 1 (TAK1) pathway in a rat model of peritoneal dialysis[J]. BMC Complement Altern Med, 2019, 19(1): 280.

[160] Zhao J L, Guo M Z, Zhu J J, et al. Curcumin suppresses epithelial-to-mesenchymal transition of peritoneal mesothelial cells (HMrSV5) through regulation of transforming growth factor-activated kinase 1 (TAK1)[J]. Cell Mol Biol Lett, 2019, 24: 32.

[161] Awaad A, Abdel Aziz H O. Iron biodistribution profile changes in the rat spleen after administration of high-fat diet or iron supplementation and the role of curcumin[J]. J Mol Histol, 2021, 52(4): 751-766.

[162] Krupa P, Svobodova B, Dubisova J, et al. Nano-formulated curcumin (Lipodisq^{TM}) modulates the local inflammatory response, reduces glial scar and preservesthe white matter after spinal cord injury in rats[J]. Neuropharmacology, 2019, 155: 54-64.

[163] Chaudhary N, Ueno-Shuto K, Ono T, et al. Curcumin down-regulates toll-like receptor-2 gene expression and function in human cystic fibrosis bronchial epithelial cells[J]. Biol Pharm Bull, 2019, 42(3): 489-495.

[164] Fan J Y, Wang Q, Zhang Z Y, et al. Curcumin mitigates the epithelial-to-mesenchymal transition in biliary epithelial cells through upregulating CD109 expression[J]. Drug Dev Res, 2019, 80(7): 992-999.

[165] Kim J M, Kim J W, Choi M E, et al. Protective effects of curcumin on radioiodine-induced salivary gland dysfunction in mice[J]. J Tissue Eng Regen Med, 2019, 13(4): 674-681.

[166] Islam R, Dash D, Singh R. Intranasal curcumin and sodium butyrate modulates airway inflammation and fibrosis via HDAC inhibition in allergic asthma[J]. Cytokine, 2022, 149: 155720.

[167] Du C, Tian Y F, Duan W L, et al. Curcumin enhances the radiosensitivity of human urethral scar fibroblasts by apoptosis, cell cycle arrest and downregulation of Smad4 via autophagy[J]. Radiat Res, 2021, 195(5): 452-462.

[168] Yu W K, Hwang W L, Wang Y C, et al. Curcumin suppresses TGF-β1-induced myofibroblast differentiation and attenuates angiogenic activity of orbital fibroblasts[J]. Int J Mol Sci, 2021,

22(13): 6829.

[169] Xiao S, Deng Y H, Shen N, et al. Curc-mPEG454, a PEGylated curcumin derivative, as a multi-target anti-fibrotic prodrug[J]. Int Immunopharmacol, 2021, 101(PtA): 108166.

[170] Sudirman S, Lai C S, Yan Y L, et al. Histological evidence of chitosan-encapsulated curcumin suppresses heart and kidney damages on streptozotocin-induced type-1 diabetes in mice model[J]. Sci Rep, 2019, 9(1): 15233.

[171] Nguyen V Q, You D G, Kim C H, et al. An anti-DR5 antibody-curcumin conjugate for the enhanced clearance of activated hepatic stellate cells[J]. Int J Biol Macromol, 2021, 192: 1231-1239.

[172] Zhang T, Li Y P, Song Y, et al. Curcumin- and cyclopamine-loaded liposomes to enhance therapeutic efficacy against hepatic fibrosis[J]. Drug Des Devel Ther, 2020, 14: 5667-5678.

[173] Yu X, Yuan L, Zhu N, et al. Fabrication of antimicrobial curcumin stabilized platinum nanoparticles and their anti-liver fibrosis activity for potential use in nursing care[J]. J Photochem Photobiol B, 2019, 195: 27-32.

[174] Hernández M, Wicz S, Pérez Caballero E, et al. Dual chemotherapy with benznidazole at suboptimal dose plus curcumin nanoparticles mitigates Trypanosoma cruzi-elicited chronic cardiomyopathy[J]. Parasitol Int, 2021, 81: 102248.

[175] Diao J Y, Wei J, Yan R, et al. Effects of resveratrol on regulation on UCP2 and cardiac function in diabetic rats[J]. J Physiol Biochem, 2019, 75(1): 39-51.

[176] Chen C, Zou L X, Lin Q Y, et al. Resveratrol as a new inhibitor of immunoproteasome prevents PTEN degradation and attenuates cardiac hypertrophy after pressure overload[J]. Redox Biol, 2019, 20: 390-401.

[177] Li P P, Song X L, Zhang D W, et al. Resveratrol improves left ventricular remodeling in chronic kidney disease via Sirt1-mediated regulation of FoxO1 activity and MnSOD expression[J]. Biofactors, 2020, 46(1): 168-179.

[178] Lieben Louis X, Meikle Z, Chan L, et al. Divergent effects of resveratrol on rat cardiac fibroblasts and cardiomyocytes[J]. Molecules, 2019, 24(14): 2604.

[179] Chen T S, Chuang S Y, Shen C Y, et al. Antioxidant Sirt1/Akt axis expression in resveratrol pretreated adipose-derived stem cells increases regenerative capability in a rat model with cardiomyopathy induced by diabetes mellitus[J]. J Cell Physiol, 2021, 236(6): 4290-4302.

[180] ShamsEldeen A M, Ashour H, Shoukry H S, et al. Combined treatment with systemic resveratrol and resveratrol preconditioned mesenchymal stem cells, maximizes antifibrotic action in diabetic cardiomyopathy[J]. J Cell Physiol, 2019, 234(7): 10942-10963.

[181] Dehghani A, Hafizibarjin Z, Najjari R, et al. Resveratrol and 1,25-dihydroxyvitamin D co-administration protects the heart against D-galactose-induced aging in rats: evaluation of serum

and cardiac levels of klotho[J]. Aging Clin Exp Res, 2019; 31(9): 1195-1205.

[182] Li Y F, Feng L F, Li G R, et al. Resveratrol prevents ISO-induced myocardial remodeling associated with regulating polarization of macrophagesthrough VEGF-B/AMPK/NF-kB pathway [J]. Int Immunopharmacol, 2020, 84: 106508.

[183] Feng H, Mou S Q, Li W J, et al. Resveratrol inhibits ischemia-induced myocardial senescence signals and NLRP3 inflammasome activation[J]. Oxid Med Cell Longev, 2020, 2020: 2647807.

[184] Raj P, Sayfee K, Parikh M, et al. Comparative and combinatorial effects of resveratrol and sacubitril/valsartan alongside valsartan on cardiac remodeling and dysfunction in MI-induced rats [J]. Molecules, 2021, 26(16): 5006.

[185] Chen Y, He T, Zhang Z J, et al. Activation of SIRT1 by resveratrol alleviates pressure overload-induced cardiac hypertrophy via suppression of TGF-β1 signaling[J]. Pharmacology, 2021, 106 (11-12): 667-681.

[186] Zou L X, Chen C, Yan X, et al. Resveratrol attenuates pressure overload-induced cardiac fibrosis and diastolic dysfunction via PTEN/AKT/Smad2/3 and NF-κB signaling pathways[J]. Mol Nutr Food Res, 2019, 63(24): e1900418.

[187] Zaparina O, Rakhmetova A S, Kolosova N G, et al. Antioxidants resveratrol and SkQ1 attenuate praziquantel adverse effects on the liver in *Opisthorchis felineus* infected hamsters[J]. Acta Trop, 2021, 220: 105954.

[188] Zheng Y, Hu G D, Wu W, et al. Transcriptome analysis of juvenile genetically improved farmed tilapia (*Oreochromis niloticus*) livers by dietary resveratrol supplementation[J]. Comp Biochem Physiol C Toxicol Pharmacol, 2019, 223: 1-8.

[189] Chen T S, Ju D T, Day C H, et al. Protective effect of autologous transplantation of resveratrol preconditioned adipose-derived stem cells in the treatment of diabetic liver dysfunction in rat model[J]. J Tissue Eng Regen Med, 2019, 13(9): 1629-1640.

[190] Cano-Martínez A, Bautista-Pérez R, Castrejón-Téllez V, et al. Resveratrol and quercetin as regulators of inflammatory and purinergic receptors to attenuate liver damage associated to metabolic syndrome[J]. Int J Mol Sci, 2021, 22(16): 8939.

[191] de Oliveira C M, Martins L A M, de Sousa A C, et al. Resveratrol increases the activation markers and changes the release of inflammatory cytokines of hepatic stellate cells[J]. Mol Cell Biochem, 2021, 476(2): 649-661.

[192] Mohseni R, Arab Sadeghabadi Z, Goodarzi M T, et al. Co-administration of resveratrol and beta-aminopropionitrile attenuates liver fibrosis development via targeting lysyl oxidase in CCl_4-induced liver fibrosis in rats[J]. Immunopharmacol Immunotoxicol, 2019, 41(6): 644-651.

[193] Li C X, Zhang R R, Zhan Y T, et al. Resveratrol inhibits hepatic stellate cell activation via the Hippo pathway[J]. Mediators Inflamm, 2021, 2021: 3399357.

[194] Yu B, Qin S Y, Hu B L, et al. Resveratrol improves CCl$_4$-induced liver fibrosis in mouse by upregulating endogenous IL-10 to reprogramme macrophages phenotype from M(LPS) to M(IL-4)[J]. Biomed Pharmacother, 2019, 117: 109110.

[195] Li S Y, Zheng X Y, Zhang X Y, et al. Exploring the liver fibrosis induced by deltamethrin exposure in quails and elucidating the protective mechanism of resveratrol[J]. Ecotoxicol Environ Saf, 2021, 207: 111501.

[196] ShamsEldeen A M, Al-Ani B, Ebrahim H A, et al. Resveratrol suppresses cholestasis-induced liver injury and fibrosis in rats associated with the inhibition of TGFβ1-Smad3-miR21 axis and profibrogenic and hepatic injury biomarkers[J]. Clin Exp Pharmacol Physiol, 2021, 48(10): 1402-1411.

[197] Chen T T, Peng S, Wang Y, et al. Improvement of mitochondrial activity and fibrosis by resveratrol treatment in mice with schistosoma japonicum infection[J]. Biomolecules, 2019, 9(11): 658.

[198] Zhu L L, Mou Q J, Wang Y H, et al. Resveratrol contributes to the inhibition of liver fibrosis by inducing autophagy via the microRNA-20a-mediated activation of the PTEN/PI3K/AKT signaling pathway[J]. Int J Mol Med, 2020, 46(6): 2035-2046.

[199] Ding S B, Wang H F, Wang M R, et al. Resveratrol alleviates chronic "real-world" ambient particulate matter-induced lung inflammation and fibrosis by inhibiting NLRP3 inflammasome activation in mice[J]. Ecotoxicol Environ Saf, 2019, 182: 109425.

[200] Wang Z Y, Li X N, Chen H, et al. Resveratrol alleviates bleomycin-induced pulmonary fibrosis via suppressing HIF-1α and NF-κB expression[J]. Aging (Albany NY), 2021, 13(3): 4605-4616.

[201] Azmoonfar R, Amini P, Yahyapour R, et al. Mitigation of radiation-induced pneumonitis and lung fibrosis using alpha-lipoic acid and resveratrol[J]. Antiinflamm Antiallergy Agents Med Chem, 2020, 19(2): 149-157.

[202] Peng X, Su H Y, Liang D L, et al. Ramipril and resveratrol co-treatment attenuates RhoA/ROCK pathway-regulated early-stage diabetic nephropathy-associated glomerulosclerosis in streptozotocin-induced diabetic rats[J]. Environ Toxicol, 2019, 34(7): 861-868.

[203] Liu S Y, Zhao M, Zhou Y J, et al. Resveratrol exerts dose-dependent anti-fibrotic or pro-fibrotic effects in kidneys: A potential risk to individuals with impaired kidney function[J]. Phytomedicine, 2019, 57: 223-235.

[204] Chen C C, Chang Z Y, Tsai F J, et al. Resveratrol pretreatment ameliorates concanavalin A-induced advanced renal glomerulosclerosis in aged mice through upregulation of Sirtuin 1-mediated klotho expression[J]. Int J Mol Sci, 2020, 21(18): 6766.

[205] Beshay O N, Ewees M G, Abdel-Bakky M S, et al. Resveratrol reduces gentamicin-induced

EMT in the kidney via inhibition of reactive oxygen species andinvolving TGF-β/Smad pathway [J]. Life Sci, 2020, 258: 118178.

[206] Feng S D, Wang J J, Teng J, et al. Resveratrol plays protective roles on kidney of uremic rats via activating HSP70 expression[J]. Biomed Res Int, 2020, 2020: 2126748.

[207] Vicari E, Arancio A, Catania V E, et al. Resveratrol reduces inflammation-related prostate fibrosis[J]. Int J Med Sci, 2020, 17(13): 1864-1870.

[208] Tang Z M, Ding J C, Zhai X X. Effects of resveratrol on the expression of molecules related to the mTOR signaling pathway in pathological scar fibroblasts[J]. G Ital Dermatol Venereol, 2020, 155(2): 161-167.

[209] Pang K, Li B B, Tang Z M, et al. Resveratrol inhibits hypertrophic scars formation by activating autophagy via the miR-4654/Rheb axis[J]. Mol Med Rep, 2020, 22(4): 3440-3452.

[210] Si L B, Zhang M Z, Guan E L, et al. Resveratrol inhibits proliferation and promotes apoptosis of keloid fibroblasts by targeting HIF-1α[J]. J Plast Surg Hand Surg, 2020, 54(5): 290-296.

[211] Yao Q C, Wu Q C, Xu X Y, et al. Resveratrol ameliorates systemic sclerosis via suppression of fibrosis and inflammation through activation of SIRT1/mTOR signaling[J]. Drug Des Devel Ther, 2020, 14: 5337-5348.

[212] Smith A J O, Eldred J A, Wormstone I M. Resveratrol inhibits wound healing and lens fibrosis: A putative candidate for posterior capsule opacification prevention[J]. Invest Ophthalmol Vis Sci, 2019, 60(12): 3863-3877.

[213] Cho D Y, Zhang S Y, Lazrak A, et al. Resveratrol and ivacaftor are additive G551D CFTR-channel potentiators: Therapeutic implications for cystic fibrosis sinus disease[J]. Int Forum Allergy Rhinol, 2019, 9(1): 100-105.

[214] Chen T S, Kuo C H, Day C H, et al. Resveratrol increases stem cell function in the treatment of damaged pancreas[J]. J Cell Physiol, 2019, 234(11): 20443-20452.

[215] Xiao Y, Qin T, Sun L K, et al. Resveratrol ameliorates the malignant progression of pancreatic cancer by inhibiting Hypoxia-induced pancreatic stellate cell activation[J]. Cell Transplant, 2020, 29: 963689720929987.

[216] Yang G X, Lyu L, Wang X H, et al. Systemic treatment with resveratrol alleviates adjuvant arthritis-interstitial lung disease in rats via modulationof JAK/STAT/RANKL signaling pathway [J]. Pulm Pharmacol Ther, 2019, 56: 69-74.

[217] Lu B Y, Corey D A, Kelley T J. Resveratrol restores intracellular transport in cystic fibrosis epithelial cells[J]. Am J Physiol Lung Cell Mol Physiol, 2020, 318(6): L1145-L1157.

[218] Vairappan B, Sundhar M, Srinivas B H. Resveratrol restores neuronal tight junction proteins through correction of ammonia and inflammation in CCl_4-induced cirrhotic mice[J]. Mol Neurobiol, 2019, 56(7): 4718-4729.

[219] Wu B B, Xiao Z X, Zhang W X, et al. A novel resveratrol-curcumin hybrid, a19, attenuates high fat diet-induced nonalcoholic fatty liver disease[J]. Biomed Pharmacother, 2019, 110: 951-960.

[220] Soussi D, Danion J, Baulier E, et al. Vectisol formulation enhances solubility of resveratrol and brings its benefits to kidney transplantation in a preclinical porcine model[J]. Int J Mol Sci, 2019, 20(9): 2268.

[221] Ančić D, Oršolić N, Odeh D, et al. Resveratrol and its nanocrystals: A promising approach for cancer therapy? [J]. Toxicol Appl Pharmacol, 2022, 435: 115851.

[222] Zhang J B, Jin H L, Feng X Y, et al. The combination of *Lonicerae Japonicae Flos* and *Forsythiae Fructus* herb-pair alleviated inflammation in liver fibrosis[J]. Front Pharmacol, 2022, 435(15): 115851.

附　　录

英文缩略词对照表

缩略词	英文全称	中文全称
AA	aristolochic acid	马兜铃酸
AAC	abdominal aortic constriction	腹主动脉缩窄
AAF	acetylaminofluorene	乙酰氨基芴
ACLT	anterior cruciate ligament transection	前交叉韧带横断
ADM	adriamycin	阿霉素
ADSCs	adipose-derived stem cells	脂肪干细胞
AhR	aryl hydrocarbon receptor	芳香烃受体
Akt	protein kinase B	蛋白激酶 B
ALB	albumin	白蛋白
ALBI	albumin-bilirubin	白蛋白-胆红素
ALK5	activin receptor-like kinase 5	激活素受体样激酶 5
ALP	serum alkaline phosphatase	血清碱性磷酸酶
ALT	alanine aminotransferase	丙氨酸转氨酶
AMPK	AMP-activated protein kinase	腺苷酸活化蛋白激酶
AMs	alveolar macrophages	肺泡巨噬细胞
Ang Ⅱ	angiotensin Ⅱ	血管紧张素 Ⅱ
AOX1	aldehyde oxidase 1	醛氧化酶 1
APOA1	apolipoprotein A 1	载脂蛋白 A1
Arg Ⅰ	arginase Ⅰ	精氨酸酶 Ⅰ
Ars	arsenic trioxide	三氧化二砷
ASC	apoptosis related spotted protein	凋亡相关斑点样蛋白

续表

缩略词	英文全称	中文全称
AS-IV	astragaloside IV	黄芪甲苷
AST	aspartate aminotransferase	天冬氨酸转氨酶
ATF4	activated transcription factor 4	激活转录因子4
ATG5	autophagy-related gene 5	自噬相关基因5
ATG7	autophagy-related gene 7	自噬相关基因7
ATP	adenosine triphosphate	腺苷三磷酸
BALF	bronchoalveolar lavage fluid	支气管肺泡灌洗液
BAPN	β-aminopropionitrile	β-氨基丙腈
Bax	Bcl-2-associated X protein	Bcl-2相关X蛋白
Bcl-2	B-cell lymphoma-2	B淋巴细胞瘤-2
BCNI	bilateral cavernous nerve injury	双侧海绵体神经损伤
BDL	bile duct ligation	胆管结扎
Beclin1	mammalian ortholog of yeast ATG 6	酵母ATG 6同系物
BECs	biliary epithelial cells	胆管上皮细胞
bFGF	basic fibroblast growth factor	碱性成纤维细胞生长因子
BLM	bleomycin	博来霉素
BMP-7	bone morphogenetic protein-7	骨形态发生蛋白-7
BPA	bisphenol A	双酚A
Bsg	basigin	基质金属蛋白酶诱导因子
BUN	blood urea nitrogen	血尿素氮
BZ	benzonidazole	苯硝唑
CAMK II	calmodulin-dependent protein kinase II	钙调蛋白依赖蛋白激酶II
cAMP	cyclic Adenosine monophosphate	环磷酸腺苷
CaOx	Ca-oxalate	草酸钙
Caspase-1	cysteinyl aspartate specific proteinase-1	半胱氨酸-天冬氨酸蛋白酶-1
CAT	catalase	过氧化氢酶
Cat B	cathepsins B	组织蛋白酶B
Cav-1	caveolin-1	小窝蛋白1
CB1	cannabinoid receptor 1	大麻素受体1
CB2	cannabinoid receptor 2	大麻素受体2

续表

缩略词	英文全称	中文全称
CCL12	CC motif chemokine ligand 12	CC 基序趋化因子配体 12
CCL2	CC chemokine ligand 2	CC 趋化因子配体 2
CCl_4	carbon tetrachloride	四氯化碳
CENPF	centromere protein F	着丝粒蛋白 F
CF	cardiac fibroblast	心脏成纤维细胞
CFTR	cystic fibrosis transmembrane conductance regulator	囊性纤维化跨膜电导调节因子
CIH	chronic intermittent hypoxia	慢性间歇性缺氧
CKD	chronic kidney disease	慢性肾脏病
COL-1	collagen 1	胶原蛋白-1
Con A	concanavalin A	伴刀豆球蛋白 A
COX-2	cyclooxygenase 2	环氧合酶 2
CRF	chronic renal failure	慢性肾功能衰竭
CTGF	connective tissue growth factor	结缔组织生长因子
Cx40	connexin 40	缝隙连接蛋白 40
CyclinD1	cell cycle protein D1	细胞周期蛋白 D1
CYP 26A1	cytochrome P450 26A1	细胞色素 P450 家族 26 亚家族 A 成员 1
CyPA	cyclophilin A	亲环素 A
DAG	diacylglycerol	甘油二酯
DB	direct bilirubin	直接胆红素
DCC	antibody-curcumin conjugate	抗体-姜黄素偶联物
DCM	dilated cardiomyopathy	糖尿病心肌病
DCs	dendritic cells	树突状细胞
DDAH1	dimethylarginine dimethylamino hydrolase 1	二甲基精氨酸二甲胺水解酶 1
DDC	3,5-diethoxycarboxyl-1,4-dihydro-2,4,6-trimethylpyridine	3,5-二乙氧基羰基-1,4-二氢-2,4,6-三甲基吡啶
DMN	dimethylnitrosamine	二甲基亚硝胺
DN	diabetic nephropathy	糖尿病肾病
DOCA	deoxycortone acetate	醋酸去氧皮质酮

续表

缩略词	英文全称	中文全称
ADM	doxorubicin	阿霉素
DR5	death receptor 5	死亡受体5
E-cadherin	epithelial cadherin	上皮钙黏蛋白
ECM	extracellular matrix	细胞外基质
EF	epidural fibrosis	硬膜外纤维化
EGFR	epidermal growth factor receptor	表皮生长因子受体
eIF2α	eukaryotic initiation factor 2α	真核细胞起始因子2α
EMT	epithelial-mesenchymal transition	上皮间质转化
EndMT	endothelial-to-mesenchymal transition	内皮间质转化
eNOS	endothelial nitric oxide synthetase	内皮型一氧化氮合酶
ER	endoplasmic reticulum	内质网
ERK1	extracellular signal-regulated kinase 1	胞外信号调节激酶1
ERα	estrogen receptor α	雌激素受体α
FA	folic acid	叶酸
FAK	focal adhesion kinase	黏着斑激酶
Fas	factor associated suicide	凋亡相关因子
FFA	free fatty acids	游离脂肪酸
FN	fibronectin	纤连蛋白
FOXO3a	forkhead box 3a	叉头框转录因子3a
FXR	farnesoid X receptor	法尼醇X受体
GATA4	GATA-binding protein 4	GATA结合蛋白4
GCLC	glutamate cysteine ligase catalytic subunit	谷氨酸-半胱氨酸连接酶催化亚基
GCLM	glutamate cysteine ligase modifier subunit	谷氨酸-半胱氨酸连接酶修饰亚基
GEN	gentamicin	庆大霉素
Gli-1	glioma associated oncogene homolog 1	胶质瘤相关癌基因-1
TAO	thyroid-associated ophthalmopathy	甲状腺相关性眼病
G-Rb1	ginsenoside Rb1	人参皂苷Rb1
G-Rg3	ginsenoside Rg3	人参皂苷Rg3
GPX4	glutathione peroxidase 4	谷胱甘肽过氧化物酶-4
GRP78	glucose regulated protein 78	葡萄糖调节蛋白78

续表

缩略词	英文全称	中文全称
GSH-Px	glutathione peroxidase	谷胱甘肽过氧化物酶
H_2O_2	hydrogen peroxide	过氧化氢
HA	hyaluronan	透明质酸
HBE	human bronchial epithelial	人支气管上皮
Hcy	homocysteine	同型半胱氨酸
HDAC1	histone deacetylasel	组蛋白去乙酰化酶1
HDL	high density lipoprotein	高密度脂蛋白
HFD	high-fat diet	高脂饮食
HG	high glucose	高糖
Hh	hedgehog signaling pathway	刺猬信号通路
$HgCl_2$	mercuric chloride	氯化汞
HIF-1	Hypoxia-inducible factor-1	缺氧诱导因子-1
HIF-1α	hypoxia inducible factor-1α	低氧诱导因子-1α
HK-2	human proximal tubular epithelial cells	人近端肾小管上皮细胞
HLF	human lung fibroblasts	人肺成纤维细胞
HMC	human mesangial cells	人肾小球系膜细胞
HMGB1	high mobility group protein 1	高迁移率族蛋白1
HMrSV 5	human peritoneal mesothelial cell	人腹膜间皮细胞
HO-1	heme oxygenase-1	血红素氧合酶-1
HSCs	hepatic stellate cells	肝星状细胞
HSP70	heat shock protein 70	热休克蛋白70
HUSFs	human urethral stricture fibroblasts	人尿道瘢痕成纤维细胞
HUVEC	human umbilical vein endothelial cell	人脐静脉内皮细胞
HYP	hydroxyproline	羟脯氨酸
ICA	icariin	淫羊藿苷
IDO1	indoleamine-2,3-dioxygenase 1	吲哚胺-2,3-双加氧酶1
IGF1	insulin-like growth factor 1	胰岛素样生长因子1
IL-1β	interleukin-1β	白介素-1β
IL-6	interleukin-6	白介素-6
iNOS	inducible nitric oxide synthase	诱导型一氧化氮合酶

续表

缩略词	英文全称	中文全称
IPF	idiopathic pulmonary interstitial fibrosis	特发性肺间质纤维化
IPSS	international prostate symptom score	国际前列腺症状评分
IR	ischemia-reperfusion	缺血再灌注
IRS1	insulin receptor substrate 1	胰岛素受体底物1
ISO	isoproterenol	异丙肾上腺素
IV-C	type IV collagen	IV型胶原
JNK	c-Jun N-terminal kinase	c-Jun氨基端蛋白激酶
Ki67	cell proliferation-associated nuclear antigen Ki67	细胞增殖相关抗原
KIF11	kinesin family number 11	驱动蛋白家族成员11
KLF4	krüpple-like factor 4	Krüpple样因子4
LC3	microtubule associated protein 1 light chain 3	微管相关蛋白1轻链3
LD	lipid droplets	脂滴
LDH	lactate dehydrogenase	乳酸脱氢酶
LF	liver fibrosis	肝纤维化
L-FABP	liver-type fatty acid-binding protein	肝脏型脂肪酸结合蛋白
LMSCs	lung mesenchymal stem cells	肺间充质干细胞
LN	laminin	层粘连蛋白
LOX	lipoxygenase	脂氧合酶
LPS	lipopolysaccharide	脂多糖
Lut	luteolin	木犀草素
LV	left ventricle	左心室
LX-2	human hepatic stellate cell line	人肝星状细胞系
MAPK	mitogen-activated protein kinase	丝裂原活化蛋白激酶
MAT2A	methionine adenosyl transferase	甲硫氨酸腺苷转移酶2A
MCD	methionine-choline deficient	蛋氨酸/胆碱缺乏
MCP-1	monocyte chemoattractant protein-1	单核细胞趋化蛋白-1
MDA	malondialdehyde	丙二醛
ESLD	end-stage liver disease	终末期肝病
MF	myofibroblast	肌成纤维细胞

续表

缩略词	英文全称	中文全称
MI	myocardial infarction	心肌梗死
miR-145	MicroRNA-145	小 RNA-145
MMP	matrix metalloproteinase	基质金属蛋白酶
MPO	myeloperoxidase	髓过氧化物酶
Mrc1	mannose receptor C1	甘露糖受体 C1 基因
MS	metabolic syndrome	代谢综合征
mTOR	mammalian target of rapamycin	哺乳动物雷帕霉素靶蛋白
MyD88	myeloid differentiation protein 88	髓样分化蛋白 88
NAFLD	nonalcoholic fatty liver disease	非酒精性脂肪肝病
NASH	nonalcoholic steatohepatitis	非酒精性脂肪性肝炎
NF-κB	nuclear factor-kappa B	核因子-κB
NIH/3T3	mouse embryonic fibroblasts	小鼠胚胎成纤维细胞
NIH-CPSI	National Institutes of Health chronic prostatitis symptom index	美国国立卫生研究院慢性前列腺症状指数
NLR	nucleotide-binding oligomerization domain-like receptors	核苷酸结合寡聚化结构域样受体
NLRP3	NOD-like receptor protein 3	NOD 样受体蛋白 3
NO	nitric oxide	一氧化氮
NOX1	NADPH oxidase 1	NADPH 氧化酶 1
NOX2	NADPH oxidase 2	NADPH 氧化酶 2
NOX4	NADPH oxidase 4	NADPH 氧化酶 4
Nrf2	nuclear factor erythroid 2-related factor 2	核因子红系 2 相关因子 2
NRK-52E	renal tubular epithelial cells	肾小管上皮细胞
OA	osteoarthritis	骨关节炎
OPN	osteopontin	骨桥蛋白
OSMF	oral submucous fibrosis	口腔黏膜下纤维化
p21waf1	cell cycle regulatory protein	细胞周期调控蛋白
P2X7r	P2X7 purinergic receptor	嘌呤能配体门控离子通道 7 受体
P53	tumor protein 53	肿瘤蛋白 53
p70S6K	70 kDa ribosomal protein S6 kinase	70 kDa 核糖体蛋白 S6 激酶

续表

缩略词	英文全称	中文全称
PAI	plasminogen activator inhibitor	纤溶酶原激活抑制剂
Parkin	Human Parkinson disease protein 2	人帕金森蛋白2
PCⅢ	Type Ⅲ procollagen	血清Ⅲ型前胶原
PCNA	proliferating cell nuclear antigen	增殖细胞核抗原
PD	peritoneal dialysis	腹膜透析
PDGFA	platelet-derived growth factor subunit A	血小板衍生生长因子A
PDGFs	platelet-derived growth factors	血小板衍生生长因子
PERK	protein kinase R-like endoplasmic reticulum kinase	内质网应激蛋白激酶R样内质网激酶
PES	prostate secretion	前列腺分泌
PF	prostate fibrosis	前列腺纤维化
PGC-1α	peroxisome proliferator-activated receptor gamma coactivator-1α	过氧化物酶体增殖活化受体γ共激活因子1-α
PGE2	prostaglandin E 2	前列环素E2
PHD	prolyl hydroxylase	脯氨酰羟化酶
PI3K	phosphatidylinositol 3-kinase	磷酸化磷脂酰肌醇-3-激酶
PⅢNP	procollagen type Ⅲ N-terminal peptide	Ⅲ型胶原的N端前肽
PINK 1	PTEN induced putative kinase 1	PTEN诱导激酶1
PⅠNP	procollagen type Ⅰ N-terminal peptide	Ⅰ型胶原N-端前肽
PIP_2	phosphatidylinositol 4,5-bisphosphate	磷脂酰肌醇-4,5二磷酸
PKB	protein kinase B	蛋白激酶B
PKCα	protein kinase C α	蛋白激酶Cα
PMCs	peritoneal mesothelial cells	腹膜间皮细胞
PON1	paraoxonase1	对氧磷酶1
PPARα	peroxisome proliferator-activated receptor-α	过氧化物酶体增殖物激活受体-α
PQ	paraquat	百草枯
PSCs	pancreatic stellate cells	胰腺星状细胞
PTEN	phosphatase and tensin homolog gene	磷酸酶及张力蛋白同源基因
PTGES 2	prostaglandin E synthase 2	前列腺素E合酶2
PZQ	praziquantel	吡喹酮

缩略词	英文全称	中文全称
RA	rheumatoid arthritis	类风湿关节炎
RAC1	Ras-related C3 botulinum toxin substrate 1	RAS 相关 C3 肉毒杆菌毒素底物 1
RANKL	nuclear factor-B receptor activator	NF-κB 受体活化因子
Res	resveratrol	白藜芦醇
RGMB	repulsive guidance molecule B	排斥性引导分子 B
Rheb	recombinant human Ras homolog enriched in brain	脑中富含的 Ras 同源蛋白
RhoA	Ras homologous family member A	Ras 同源蛋白家族成员 A
RAI	radioactive iodine	放射性碘
RIF	renal interstitial fibrosis	肾间质纤维化
ROCK	Rho-associated coiled coil-forming protein kinase	Rho 相关螺旋卷曲蛋白激酶
ROS	reactive oxygen species	活性氧
RXRα	retinoic acid X receptor α	视黄酸 X 受体 α
SASP	senescence-associated secretory phenotype	衰老相关分泌表型
SCFA	short-chain fatty acid	短链脂肪酸
SCr	serum creatinine	血清肌酐
sFRP2	secreted frizzled-related protein two gene	分泌的卷曲相关蛋白 2
SG	salivary gland	唾液腺
SiO_2	silicon dioxide	二氧化硅
Sirt1	silent information regulator factor 2 related enzyme 1	沉默信息调节因子 2 相关酶 1
Smurf2	smad ubiquitination regulatory factor 2	Smad 泛素化调节因子 2
SOCS3	suppressor of cytokine signaling 3	细胞因子信号传送抑物 3
SOD	superoxide dismutase	超氧化物歧化酶
Sp1	specificity protein 1	特化蛋白 1
SREBP-2	sterol-regulatory element binding protein 2	胆固醇调节元件结合蛋白 2
SSc	systemic sclerosis	系统性硬化症
STAT3	signal transduction and transcriptional activator 3	信号转导及转录激活因子 3
STZ	streptozotocin	链脲佐菌素

续表

缩略词	英文全称	中文全称
T1DM	type 1 diabetes mellitus	1型糖尿病
T2DM	type 2 diabetes mellitus	2型糖尿病
TAA	thioacetamide	硫代乙酰胺
TAC	transverse aortic constriction	横向主动脉缩窄
TAK1	TGF-β-activated kinase 1	转化生长因子β激活激酶1
TanⅡA	tanshinone ⅡA	丹参酮ⅡA
TAZ	transcriptional coactivator with PDZ-binding motif	含PDZ结合基序的转录共激活因子
TB	total bilirubin	总胆红素
TGF-β	transforming growth factor-β	转化生长因子-β
TIMP-1	tissue inhibitor of matrix metalloproteinase 1	基质金属蛋白酶抑制剂-1
TLR4	Toll-like receptor 4	Toll样受体4
TNF-α	tumor necrosis factor-α	肿瘤坏死因子-α
TOP2A	topoisomerase 2α	拓扑异构酶2α
TP	total protein	总蛋白
TRPM7	melastatin-related transient receptor potential 7	M型瞬时受体电位通道7
TS	takotsubo cardiomyopathy	心尖球囊样
THBS-1	thrombospondin-1	血小板反应素-1
UA	serum uric acid	血清尿酸
UAN	uric acid nephropathy	尿酸肾病
UCP2	uncoupling protein 2	解偶联蛋白2
uPA	uroplasminogen activato	尿型纤溶酶原激活剂
UUO	unilateral ureteral obstruction	单侧输尿管梗阻
VCAM-1	vascular cell adhesion molecule-1	血管细胞黏附分子-1
VE-cadherin	vascular endothelial-cadherin	血管内皮钙黏蛋白
VEGF	vascular endothelial growth factor	血管内皮生长因子
WBC	white blood cell	白细胞
YAP	yes-associated protein	YES相关蛋白
α-SMA	α-smooth muscle actin	α-平滑肌蛋白
β-catenin	beta-catenin	β-连环蛋白